Sustainable Textiles: Production, Processing, Manufacturing & Chemistry

Series Editor

Subramanian Senthilkannan Muthu, Green Story Inc., Tung Chung, Hong Kong

This series aims to address all issues related to sustainability through the lifecycles of textiles from manufacturing to consumer behavior through sustainable disposal. Potential topics include but are not limited to: Environmental Footprints of Textile manufacturing; Environmental Life Cycle Assessment of Textile production; Environmental impact models of Textiles and Clothing Supply Chain; Clothing Supply Chain Sustainability; Carbon, energy and water footprints of textile products and in the clothing manufacturing chain; Functional life and reusability of textile products; Biodegradable textile products and the assessment of biodegradability; Waste management in textile industry; Pollution abatement in textile sector; Recycled textile materials and the evaluation of recycling; Consumer behavior in Sustainable Textiles; Eco-design in Clothing & Apparels; Sustainable polymers & fibers in Textiles; Sustainable waste water treatments in Textile manufacturing; Sustainable Textile Chemicals in Textile manufacturing. Innovative fibres, processes, methods and technologies for Sustainable textiles; Development of sustainable, eco-friendly textile products and processes; Environmental standards for textile industry; Modelling of environmental impacts of textile products; Green Chemistry, clean technology and their applications to textiles and clothing sector; Eco-production of Apparels, Energy and Water Efficient textiles. Sustainable Smart textiles & polymers, Sustainable Nano fibers and Textiles; Sustainable Innovations in Textile Chemistry & Manufacturing; Circular Economy, Advances in Sustainable Textiles Manufacturing; Sustainable Luxury & Craftsmanship; Zero Waste Textiles.

Subramanian Senthilkannan Muthu
Editor

Technical Textiles for Sustainability

Springer

Editor
Subramanian Senthilkannan Muthu
Green Story Inc
Toronto, Canada

ISSN 2662-7108 ISSN 2662-7116 (electronic)
Sustainable Textiles: Production, Processing, Manufacturing & Chemistry
ISBN 978-3-032-29973-4 ISBN 978-3-032-29974-1 (eBook)
https://doi.org/10.1007/978-3-032-29974-1

This Springer imprint is published by the registered company Springer Nature Switzerland AG
The registered company address is: Gewerbestrasse 11, 6330 Cham, Switzerland

This book is dedicated to:

The lotus feet of my beloved Lord
Pazhaniandavar
My beloved late Father
My beloved Mother
My beloved Wife Karpagam and
Daughters—Anu and Karthika
My beloved Brother
Last but not least
To everyone working in technical textiles
sector to make the world, "SUSTAINABLE"

Contents

Introduction to the Book: Technical Textiles for Sustainability

Subramanian Senthilkannan Muthu

Keywords Technical Textiles · Biodegradation · Sustainability · Sustainable fibers · Waste Water · Protective Textiles · Sustainable Development · Sustainable Clothing

With the growing global emphasis on environment sustainability and efficient resource utilization, there is an obvious and urgent need to study the role of technical textiles in advancing sustainable development. The rapid development of industrial and high-performance textiles has generated significant opportunities, while simultaneously rising concerns regarding various environmental impacts. This book chapters critically addresses the challenges by emphasizing the development and application of sustainable materials, eco-efficient processes and innovative solutions that align with sustainable technical textiles.

As sustainability assumes a key role in technical education, research and industrial practices, this book serves as a vital resource researchers, students, academicians and professional aiming to understand and implement sustainable solutions within the area of technical textiles. It brings together emerging research and innovative strategies for sustainable material utilization encompassing areas such as ethnobotanical perceptions, agro residue valorization, bio-based material development and circular recycling technologies. The book addresses the themes including carbon neutral applications, the utilization of cotton stalks residues, regenerative textile approaches, the development of natural and ecofriendly fibers and recent advancement in textile recycling processes. By presenting these perspectives, this book outlines viable pathways toward a climate resilient and resource efficient future.

S. S. Muthu (✉)
Green Story Inc, Toronto, Canada
e-mail: drsskannanmuthu@gmail.com

S. S. Muthu (ed.), *Technical Textiles for Sustainability*, Sustainable Textiles: Production, Processing, Manufacturing & Chemistry, https://doi.org/10.1007/978-3-032-29974-1_1

Furthermore, by addressing both the opportunities and challenges within the field, it reassures collaborative engagement among researchers, manufactures and policymakers, thereby contributing to the advancement of a sustainable textile industry.

The chapter two reports, many industries including textile, leather, printing and paper sectors extensively use synthetic dyes and their discharge into wastewater poses significant environmental challenges due to their complex chemical structures, persistence and potential toxicity. Dye contaminated effluents not only degrade water quality but also block sunlight penetration, disrupt aquatic ecosystems and exert harmful effects on human health. To address these concerns, a wide range of treatment approaches physical, chemical, biological and hybrid methods that have been developed and widely investigated for dye removal, each offering distinct advantages and limitations. In recent years advanced nanomaterials and engineered sorbents have demonstrated considerable promise in enhancing treatment efficiency. Modified adsorbents and photocatalysts such as metal dopped titanium nanotubes like W, Ag, Fe, MoS_2 doped TNTs, carbon quantum dot (CQD) doped TNTs and Cu doped Co_3O_4 have shown exceptional potential as effective adsorptive and catalytic material for dye removal, which highlights their applicability in real wastewater systems. Nanomaterials enabled treatments have achieved removal efficiencies exceeding 90–99% under optimized conditions, significantly improving dye degradation and mineralization compared to conventional methods. This chapter systematically examines current treatment strategies evaluates their performance and emphasizes the role of novel material in bridging the gap between laboratory scale research and practical implementation. Furthermore, it underscores the importance of techno-economic feasibility, long term stability, regulatory compliance and environmental sustainability in the development of dye remediation technologies. Finally, future perspectives are outlined focusing on the need for cost effective, scalable and eco-friendly solutions, integrated treatments systems and the valorization of dye laden residues for resource recovery. Collectively, these insights provide a comprehensive roadmap for advancing wastewater treatment to protect the environment and mitigate public health risks.

The third chapter stated that, the fashion industry is one of the most resource intensive sectors globally which is increasingly integrating sustainable practices across its value chain from material selection to production and distribution. Among emerging innovations, textiles derived from food waste such as orange fiber produced from citrus processing residues offer both environmental and aesthetic advantages, distinguishing them from conventional recycled materials like fibers derived from plastic bottles. This chapter examines how textile origin (Orange Fiber versus recycled plastic fiber) influences consumer willingness to purchase with particular attention to the moderating role of brand type (luxury versus fast fashion). Findings from an experimental study reveals that orange fiber elicits a higher willingness to buy compared to recycled plastic fiber in the context of luxury brands, whereas this advantage become less pronounced for fast fashion brands. Furthermore, the relationship between textile origin and purchase intention is mediated by perceived coolness, highlighting the symbolic and experimental value consumers associate with innovative sustainable materials. These results suggest that the strategic combination of novel technical textiles with strong brand positioning can significantly enhance consumer acceptance effectively aligning sustainability with perceived desirability.

The fourth chapters briefly reviews that the technical textiles represent a rapidly growing, high value segment with diverse applications in healthcare, construction, automotive manufacturing, protective equipment, sports, agriculture and packaging. These materials are engineered to deliver specialized properties such as enhanced strength, durability, flexibility, chemical resistance, flame retardancy and advanced functional finishes that are often difficult to achieve through conventional textiles. While these performance characteristics ensure reliability, they can also hinder biodegradability at the end of the product life cycle, leading to significant sustainability concerns. Achieving an optimal balance between biodegradability, technical performance, service life and cost effectiveness remains a critical challenge. This chapter reviews the biodegradability of technical textiles, with particular emphasis on the fate of their components under landfill and other disposal conditions. It examines current research on how factors such as raw material selection, manufacturing processes, chemical modifications and environmental conditions influence degradation behaviour. Furthermore, the chapter evaluates existing standards and methodologies for biodegradability testing and discusses the trade-offs between degradability and functional performance. By synthesizing current knowledge and addressing emerging challenges, this work highlights the importance of integrating environmental sustainability into the design and development of technical textiles.

The fifth chapter helps to understand the ultraviolet (UV) radiation poses a significant risk to human health, leading to eye damage, skin cancer and premature aging. These concerns have driven a growing demand for textiles that offer effective UV protection. The opening section of this chapter outlines the health impacts of UV exposure and highlights the need for protective textile solutions. Various approaches have been developed to enhance the UV resistance of textiles, including the incorporation of UV—absorbing polymers, chemical treatments and specialized surface coatings. The chapter further examines the interaction of UV radiation with textile materials through absorption, reflection and scattering and how these mechanisms are influenced by fiber type, fabric structure, coloration and finishing techniques that ultimately determining the level of protection. In addition, the durability of UV protection properties is critically evaluated, particularly their ability to withstand repeated washing, mechanical wear and environmental exposure. Standardized testing methods including those used to assess the ultraviolet protection factor (UPF) are reviewed to determine the effectiveness and long-term reliability of these protective features. Increasing emphasis is being placed on environmentally sustainable UV protective technologies. This chapter highlights the potential of bio based UV absorbents, natural dyes, eco-friendly nanoparticles and green finishing processes that minimizes chemical toxicity and environmental impact. It also underscores the importance of durability, reduced chemical leaching, recyclability and extended product lifespan as key parameters for sustainable UV-protective textiles. Collectively, these strategies contribute to reduced resource consumption and textile waste while maintaining high levels of protective performance.

The sixth chapter focuses on the fashion industry is among the most resource-intensive and environmentally impactful sectors globally, driven by high water consumption, extensive chemical usage, greenhouse gas emissions and significant textile waste generation. These challenges have been further exacerbated by fast-fashion models characterized by rapid production and consumption cycles. In response, sustainability and circular economy principles have become central to discussions on the future of fashion, with fiber innovation emerging as a key driver for reducing environmental impact. This chapter examines technical eco-innovation in textiles by focusing on sustainable and innovative fibers as a pathway toward a more circular and responsible fashion system. Drawing on existing literature in sustainable fashion, circular economy and green innovation it explores how fibers derived from waste streams and renewable biological resources can contribute to waste minimization, resource efficiency and value creation. Particular emphasis is placed on the luxury fashion sector, which faces both opportunities and inherent tensions in adopting sustainable practices while maintaining quality, heritage and exclusivity. Based on a narrative literature review and multiple case study analyses (Orange Fiber, Pinatex and Mylo) the chapter, illustrates diverse models of sustainable fiber innovation, including waste to resource conversion, plant-based leather alternatives and biotechnology driven materials. The analysis highlights both the potential and the challenges associated with scaling sustainable fibers, underscoring the importance of supply chain collaboration, technological readiness and consumer acceptance. Overall, the chapter positions sustainable technical fibers at the intersection of innovation, circularity and consumer culture contributing to a redefining the future of fashion.

I take this opportunity to thank all the contributors for their earnest efforts to bring out this book successfully. I am sure readers of this book will find it very useful.

With best wishes,

Dr. Subramanian Senthilkannan Muthu.

Sustainable Treatment of Dye-Contaminated Wastewater from Technical Textiles: Emerging Nanomaterials, Hybrid Systems, and Circular Economy Integration

Jamshid Hussain, Ndumiso Vukile Mdlovu, Mathurin François, and Kuen-Song Lin

Abstract Many industries, like textile, leather, printing, paper, and other sectors, are widely used, and their release into wastewater presents issues due to their complicated chemical compositions, persistence, and potential toxicity. Dye-contaminated effluents not only alter water quality but also block sunlight penetration, disrupt aquatic ecosystems, and have toxic effects on human health. To overcome these health and environmental issues, many approaches, such as physical, chemical, biological, and hybrid routes, have been developed and investigated for dye removal. Each one offers distinct advantages and limitations. Recently, advanced nanomaterials and tailored sorbents have demonstrated promising performance in enhancing treatment efficiency. Modified adsorbents/photocatalysts such as metal-doped titanium nanotubes (W, Ag, Fe, MoS_2-doped TNTs), CQD (carbon quantum dot)-doped TNTs, or Cu-doped Co_3O_4 have particularly shown great potential as effective adsorbents and catalytic materials for dye removal, underscoring their applicability in real wastewater scenarios. Nanomaterial-enabled treatments demonstrated >90–99% dye removal under optimized conditions, substantially improving mineralization compared to conventional systems. This chapter examines current treatment approaches in a methodical manner, evaluates their efficacy, and emphasizes the significance of novel materials in bridging the gap between lab results and practical use. Additionally, it highlights the significance of techno-economic viability, long-term durability, regulatory adherence, and environmental sustainability in the development of dye removal technologies. Finally, future directions are discussed, including the need for cost-effective, scalable, and environmentally sustainable technologies, integrated treatment systems, and valorization of dye-laden residues for resource recovery. Together, these insights provide a roadmap for advancing wastewater treatment for environmental protection and the mitigation of public health issues.

J. Hussain · N. V. Mdlovu · M. François · K.-S. Lin (✉)
Department of Chemical Engineering and Materials Science, Yuan Ze University, Chung-Li District, Taoyuan City, Taiwan
e-mail: kslin@saturn.yzu.edu.tw

S. S. Muthu (ed.), *Technical Textiles for Sustainability*, Sustainable Textiles: Production, Processing, Manufacturing & Chemistry, https://doi.org/10.1007/978-3-032-29974-1_2

Keywords Dye wastewater treatment · Environmental impacts · Human health risks · Titanium nanotubes (TNTs) · Advanced sorbents · Sustainable textile wastewater management

Abbreviations

AC	Activated carbon
AEOPs	Advanced electrochemical oxidation processes
AI	Artificial intelligence
AOPs	Advanced oxidation processes
AnSBR	Anaerobic sequencing batch reactor
BDD	Boron-doped diamond (electrode)
BOD	Biochemical oxygen demand
CAS	Conventional activated sludge
CE	Circular economy
COD	Chemical oxygen demand
COFs	Covalent organic frameworks
CNTs	Carbon nanotubes
CQDs	Carbon quantum dots
CR	Congo red
DSSCs	Dye-sensitized solar cells
DO	Dissolved oxygen
EC	Electrocoagulation
EO	Electrochemical oxidation
EPS	Extracellular polymeric substances
ETP	Effluent treatment plant
FT-IR	Fourier transform infrared spectroscopy
$g\text{-}C_3N_4$	Graphitic carbon nitride
GO	Graphene oxide
HPLC	High-performance liquid chromatography
HPLC–MS	High-performance liquid chromatography–mass spectrometry
HNTs	Halloysite nanotubes
IFAS	Integrated fixed-film activated sludge
LCA	Life cycle assessment
MB	Methylene blue
MBR	Membrane bioreactor
MBBR	Moving bed biofilm reactor
ML	Machine learning
MOFs	Metal-organic frameworks
MR	Methyl red
MS	Mass spectrometry

1 Introduction

1.1 Background and Control of Water Pollution

The economic advancement of a nation is now profoundly influenced by the extent of its industrial development. In the modern era of consumer-driven marketplaces and rapid globalization, the textile industry has become a pivotal sector, symbolizing both technological innovation and sustainable economic prosperity [1, 2]. The global textile industry, a multi-trillion-dollar enterprise, weaves together the materials that enhance daily life in garments. In many countries, like India, Sri Lanka, Vietnam, and Bangladesh, this industry contributes significantly to economic growth by creating jobs and foreign exchange, but it simultaneously presents serious environmental challenges due to insufficient regulation and oversight of chemical processing.

Current estimates suggest that around 10,000 varieties of commercial dyes are produced each year, about 700,000 tons. Of this, nearly 280,000 tons are discharged into surface water systems, causing significant ecological concerns [3, 4]. Nearly 1.6 million liters of water are used every day by a medium-sized textile mill that generates around 8000 kg of fabric, of which 16 and 8% are used for dyeing processes and printing operations, respectively. The water usage for dyeing typically varies between 30 and 50 L/kg of fabric, depending on the dye type used. For yarn dyeing, the total water utilization is likely to be around 60 L/kg of yarn, representing approximately 15–20% of the facility's total wastewater output [5]. Industrial waste, around 15–20%, is produced from textile dyeing and fabric finishing. This poses a serious and complicated environmental challenge for the textile sector [6, 7]. Textile dye effluents are therefore among the most complex and recalcitrant industrial waste streams, requiring advanced and integrated treatment strategies. Given the high pollutant loads and chemical complexity, dye wastewater is among the most challenging industrial effluents to treat.

1.2 Sources and Applications of Dyes in Various Industries

In numerous sectors, such as textiles, food, cosmetics, and printing, dyes serve as vital chemical agents for adding color to products. They are chromophoric substances capable of chemically or physically interacting with substrates, selectively absorbing specific wavelengths of light to produce visible color. Their color properties arise from conjugated π-electron systems, where electron transitions occur upon exposure to light or heat energy [8]. Food-grade dyes are usually utilized in the food industry to improve product appearance, attract consumers, and comply with regulatory standards [9]. In pharmaceuticals, dyes aid in drug formulation by enabling product differentiation, identification, and dosage accuracy. The paper and pulp industry also employs dyes to produce varied color shades for writing, printing, and packaging materials [10]. Natural dyes, sourced from plant extracts, minerals, and animals, are biodegradable, renewable, and environmentally sustainable, as demonstrated in Fig. 1. Their production generally needs fewer chemical processes, thereby minimizing pollution [11]. From a health perspective, natural dyes are largely non-toxic and hypoallergenic, making them safer for applications involving direct skin contact, such as textiles. Individuals with chemical sensitivities or allergies often prefer products colored with natural dyes. In contrast, synthetic dyes, as shown in Fig. 1, are produced from petrochemical compounds and contribute significantly to environmental contamination. Certain synthetic dyes contain hazardous substances, including heavy metals and carcinogenic compounds, which may pose health risks upon prolonged exposure or ingestion [12]. Consequently, many countries have implemented regulations to restrict the use of toxic chemicals in synthetic dyes to address these health and environmental concerns. Natural dyes create distinctive, soft, and earthy tones but generally lack the vibrant intensity achieved with synthetic dyes. Furthermore, synthetic dyes provide better color stability and brightness even after extended use, whereas natural dyes are more likely to fade when exposed to sunshine or frequent washing. This superior color fastness makes synthetic dyes more suitable for applications such as outdoor fabrics. In terms of production, natural dyes are labor-intensive and time-consuming, requiring significant manual effort [11, 12]. Conversely, synthetic dyes are more cost-effective due to the easy accessibility of raw materials and efficient manufacturing processes, making them commonly used across industries such as textiles, cosmetics, and printing. These contrasting chemical properties critically influence treatability, environmental fate, and process selection for wastewater remediation.

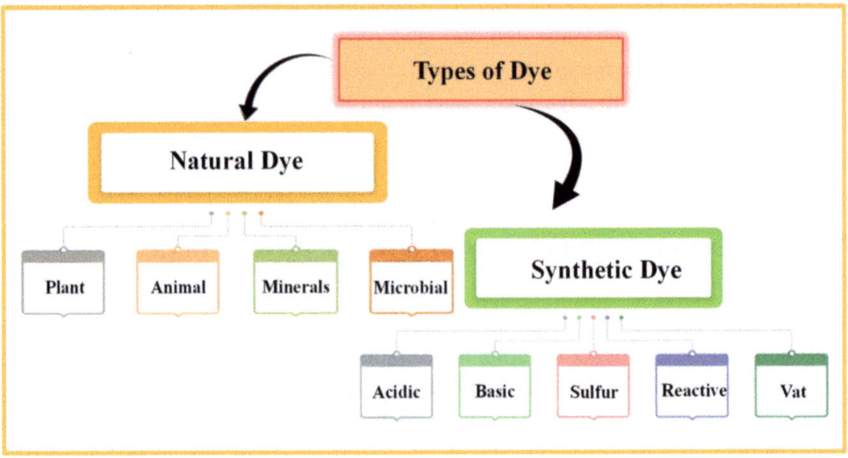

Fig. 1 Classification of dyes into natural and synthetic

1.3 Environmental and Health Impacts of Dye-Contaminated Effluents

Textile effluents are among of the most complex industrial discharges, containing highly carcinogenic substances that can cause allergic reactions and generate toxic by-products when released into the environment without prior treatment [13]. Because many dye molecules and related compounds are resistant to biodegradation and can linger in the environment for long periods of time, these untreated discharges endanger both aquatic ecosystems and people's health. A mixture of hazardous organic and inorganic compounds, such as aldehydes, ketones, tributyl phosphate, benzene derivatives, and dibutyltin, is commonly found in industrial textile effluent. These substances really add to the severe contamination of wastewater [14]. The presence of such contaminants alters the physicochemical characteristics of water bodies, raising turbidity, increasing organic load, and introducing toxic intermediates that interfere with biological self-purification processes. These pollutants degrade groundwater quality, reduce dissolved oxygen levels, disrupt aquatic ecosystems, and may indirectly contribute to climate change [15]. Moreover, such effluents are typically characterized by the presence of colloidal particles, high pollutant loads, elevated temperatures, and fluctuating levels of pH, biochemical oxygen demand (BOD), chemical oxygen demand (COD), and total dissolved solids (TDS).

1.4 Global Regulatory Landscape and Discharge Standards

Managing textile wastewater is a major topic attracting researchers' attention for ecosystem protection. It should be noted that effluents from such industrial processes can heavily burden the environment, as they contain a mixture of suspended mineral salts, oils, heavy metals and their complexes, various chemicals, dyes, and both readily and poorly biodegradable organic compounds [16]. Therefore, the use of advanced technologies to address this environmental threat has become a research priority. Zero Liquid Discharge (ZLD) technology, which emphasizes water reuse and the decrease of effluent discharge, was found in a recent thorough evaluation to be a potential method in the textile sector for sustainable wastewater management [17]. Despite its high efficiency, ZLD technology poses significant environmental challenges, including high energy consumption, a larger carbon footprint, increased greenhouse gas emissions, and the need for greater chemical usage [18–20]. Since the high energy demand and chemical use of ZLD can offset some environmental benefits and contribute to greenhouse gas emissions [61], upstream alternatives and process optimization should be considered to minimize the environmental influence of textile wastewater usage. Accordingly, environmental laws generally require the treatment of textile effluents before discharge into water bodies [21, 22]. In addition to legal enforcement, other methods such as adsorption, photocatalysis, and biological treatment offer effective alternatives to mitigate pollution from textile wastewater.

These methods help reduce toxic dyes and pollutants, support the CE, and align strongly with the SDGs. ZLD systems, while effective, impose significant energy and carbon burdens that may offset environmental benefits unless integrated with renewable sources.

1.5 UN Sustainable Development Goals (SDGs)

The SDGs of the United Nations are intimately related to the treatment of dye-contaminated wastewater because effective wastewater management promotes public health and environmental sustainability [23]. Advancing treatment technologies, especially those utilizing emerging nanomaterials, contribute significantly to SDG 6 (Clean Water and Sanitation) by improving the degradation of persistent dyes, toxic organic compounds, and heavy metals, thereby protecting freshwater resources and enabling water reuse within industrial operations [24, 25]. Given that many synthetic dyes and their degradation intermediates are carcinogenic, mutagenic, or endocrine-disrupting, efficient treatment also supports SDG 3 (Good Health and Well-Being) by controlling human exposure to harmful contaminants through drinking water, food chains, and agricultural irrigation [26]. Moreover, the development and adoption of advanced sorbents, photocatalysts, and hybrid nanomaterial-based processes contribute to SDG 9 (Industry, Innovation, and Infrastructure) by promoting cleaner production, resource-efficient manufacturing, and modern wastewater treatment systems, including decentralized and low-energy technologies suitable for small- and medium-scale enterprises [27, 28]. Dye wastewater treatment is also directly related to SDG 12 (Responsible Consumption and Production), as innovative technologies enable waste minimization, sludge reduction, resource recovery (including dye molecules, metals, and treated water), and the transition toward CE principles [29]. Because high-energy treatment systems such as ZLD can contribute to elevated carbon emissions, adopting solar-driven photocatalysis, low-temperature catalysis, and regenerable nanomaterials supports SDG 13 (Climate Action) by reducing the carbon footprint of wastewater treatment operations. Additionally, preventing the discharge of untreated dye effluents into natural water bodies directly supports SDG 14 (Life Below Water) and SDG 15 (Life on Land), as toxic dyes disrupt aquatic ecosystems, inhibit photosynthesis, degrade soil quality, and bioaccumulate in organisms [30]. Overall, sustainable dye wastewater treatment depends on effective collaboration among governments, industries, and research institutions, aligning with SDG 17 (Partnerships for the Goals) by fostering technological innovation, knowledge exchange, and capacity building [31] Thus, integrating advanced materials and sustainable treatment approaches within the textile and dyeing industries not only improves environmental and operational performance but also advances global progress toward multiple interconnected SDGs.

1.6 Scope and Objectives

This chapter presents a comprehensive and systematic examination of dye-contaminated wastewater, with particular emphasis on its origins, chemical complexity, environmental persistence, and associated ecological and human health risks. Beginning with an overview of global water pollution challenges and the substantial contribution of the textile industry to industrial wastewater discharge, the chapter delineates the sources, classifications, and physicochemical characteristics of synthetic and natural dyes. It further explores the toxicity profiles, biodegradability constraints, and analytical challenges associated with dye pollutants. The manuscript critically evaluates conventional technologies for wastewater treatment, such as physical, chemical, biological, and hybrid systems, underlining their principles, operational limitations, sludge generation issues, and performance under variable effluent conditions. In parallel, the chapter emphasizes the regulatory context and global discharge standards, underscoring the need for sustainable and energy-efficient alternatives. Building on this foundation, the scope expands to detail the emergence of nanotechnology as a transformative approach for dye removal. This work systematically reviews various classes of advanced materials, comprising metal and metal oxide nanoparticles, carbon-based nanostructures, MOFs, COFs, clay- and silica-based adsorbents, magnetic nanomaterials, and functionalized or doped hybrids. Their mechanisms of action, adsorption, photocatalysis, redox transformation, and synergistic hybrid pathways are critically analyzed alongside kinetic and isotherm models that describe dye nanomaterial interactions. The chapter further examines techno-economic considerations, including cost–benefit trade-offs, scalability constraints, operational energy demands, and environmental footprints assessed through life-cycle analysis (LCA). Resource recovery pathways, valorization of dye-laden residues, and CE integration are also addressed to contextualize the sustainability dimension of emerging technologies.

Finally, the scope encompasses forward-looking perspectives that identify current research gaps and future directions. These include creating recyclable, multipurpose nanocomposites; utilizing machine learning (ML) and artificial intelligence (AI) to optimize processes in real time; advancing of green nanomaterial synthesis routes; developing of policy and safety frameworks for nanotechnology deployment; and translating of laboratory findings into pilot-scale and industrial applications. By integrating scientific, technological, environmental, economic, and regulatory dimensions, this chapter aims to provide an authoritative reference on the state-of-the-art in dye wastewater treatment, offering insights that inform both academic research and industrial practice. The scope integrates scientific, engineering, economic, and sustainability perspectives to guide advanced dye wastewater treatment research.

1.7 Roadmap of the Chapter

A comprehensive overview is offered in this chapter for sustainable strategies for treating dye-contaminated wastewater originating from technical textiles. It begins by establishing the context and significance of the issue, providing a critical understanding of the challenges to the environment posed by textile wastewater. Following this, the chapter presents a thorough review of existing wastewater treatment technologies, analyzing their effectiveness and limitations. The discussion then transitions to emerging nanomaterials and hybrid systems, highlighting their transformative potential in enhancing treatment efficiency and promoting sustainability in wastewater management. The chapter concludes by examining the integration of circular economy principles, offering actionable insights and forward-thinking solutions for scalable and environmentally sustainable wastewater treatment strategies in the textile industry.

2 Dye Chemistry and Wastewater Characteristics

2.1 Classification of Dyes: Azo, Anthraquinone, Sulfur, Vat, Reactive, Basic

As illustrated in Table 1, dye classification based on their application method is the most practical system in the textile industry, as it links dye chemistry to the substrate and the dyeing process. Dyes are grouped according to how they interact with fibers through covalent bonding, ionic attraction, adsorption, or redox reactions, and their compatibility with specific materials such as cotton, wool, silk, polyester, or acrylic [32, 33]. For example, reactive dyes form covalent bonds with cellulose and protein fibers, providing excellent wash fastness. Vat dyes like indigo are reduced to soluble leuco forms before oxidation within cotton fibers. Disperse dyes color hydrophobic fibers like polyester by diffusion, while acid dyes fix to wool, silk, or nylon in acidic conditions. Basic dyes attach to acrylics and leather through electrostatic forces; mordant dyes require metallic ions for fixation; and sulfur dyes produce deep shades on cotton under alkaline reducing conditions [34]. Solvent dyes dissolve in organic media for plastics, whereas pigments are insoluble colorants needing binders rather than fiber affinity [35]. This classification effectively connects the chemical behavior of dyes with their industrial applications, providing a framework widely adopted in textile processing and environmental regulation [33].

Table 1 Characteristics of different dyes based on applications

Application class	Fiber type	Example dye	Bonding
Direct dyes	Cotton, linen (cellulosics)	Direct blue 1	Hydrogen bonding
Reactive dyes	Cotton, wool, silk	Reactive red 120	Covalent bond
Vat dyes	Cotton	Indigo	Covalent and hydrogen bonding
Disperse dyes	Polyester, acetate	Disperse orange 3	Physical (no chemical bond)
Acid dyes	Wool, silk, nylon	Acid red 18	Ionic bond
Basic (cationic) dyes	Acrylic, paper, leather	Basic violet 10	Ionic bond
Mordant dyes	Wool, leather	Alizarin	Coordination (metal–ligand)
Sulfur dyes	Cotton	Sulfur black 1	Covalent bond (after oxidation)
Solvent dyes	Plastics, petroleum products	Solvent yellow 14	No chemical bond
Pigments (not dyes)	Any (used with binder)	Pigment blue 15 (Cu-phthalocyanine)	Mechanical bond (via binder)

2.2 Physicochemical Properties Influencing Treatment (Solubility, Stability, Charge)

The physicochemical characteristics of dyes, particularly solubility, stability, and charge, govern their behavior in wastewater and determine the efficiency of treatment processes.

Solubility: The solubility of dyes governs how they partition in aqueous systems, directly influencing the selection of treatment methods. Water-soluble dyes, such as reactive, acid, and direct dyes, remain predominantly in the aqueous phase and exhibit low removal efficiency through conventional solid–liquid separation methods, typically achieving only 20–40% removal by sedimentation and filtration. For effective removal, these dyes usually need sophisticated treatment techniques such as membrane filtering or activated carbon adsorption. In contrast, hydrophobic dyes, like disperse and vat dyes, form fine particulate suspensions stabilized by dispersing agents. These dyes typically exhibit better removal rates in procedures such as coagulation-flocculation, dissolved-air flotation, or membrane filtration, where the removal efficiency can exceed 80–90% for some systems. The direct correlation between solubility and phase behavior and the selection of appropriate treatment processes [36, 37].

Electrolyte Effects and Ionic Strength: In processes like cotton dyeing, large quantities of electrolytes (typically 50–80 g/L NaCl or Na_2SO_4) are added to promote the exhaustion of reactive dyes. These electrolytes can significantly affect the treatment process, with ionic strengths influencing aggregation, adsorption, and electrostatic behavior of the dye molecules. Studies have shown that higher ionic strengths (such as 100 g/L NaCl) can reduce the effectiveness of coagulants by up to 30% by screening charge interactions, thus lowering coagulation performance [37, 38].

Charge/pH-Dependent Speciation: The charge behavior of dyes varies with pH. For example, reactive and acid dyes are generally anionic under the alkaline conditions used for dye fixation, while basic dyes are cationic. The ionic character of dyes is crucial when selecting suitable adsorbents and coagulants. Cationic polymers are highly effective for removing anionic dyes and can improve adsorption efficiencies by up to 50% compared to conventional adsorbents. Additionally, pH adjustment can modify the speciation of dye molecules, enhancing treatment efficiency. For instance, acidic conditions (pH < 6) can help improve the removal of basic dyes, whereas alkaline conditions (pH > 7) are more effective for acid dyes [36, 39].

Chemical Stability and Need for Advanced/Hybrid Treatment: The chemical stability of dyes, particularly those with azo, anthraquinone, and phthalocyanine chromophoric groups, leads to their persistence in wastewater. These groups are very hardy to light and microbial degradation, contributing to the high recalcitrance of these dyes in the environment. Effective treatment strategies often combine physicochemical removal with oxidative polishing methods like ozonation, persulfate/UV oxidation, Fenton oxidation, or electro-oxidation. These methods can reduce dye concentration by 90–95% in a relatively short time. Recent studies also show that matrix ions (e.g., 20 g/L NaCl) can modulate radical pathways, enhancing sulfate/$^{\cdot}$OH oxidation systems and improving the degradation efficiency by up to 40% [36, 40].

Temperature and auxiliaries: Dye baths in textile processing are typically warm, with temperatures ranging from 40 to 70 °C. Elevated temperatures lower the viscosity of the dye solution, thereby enhancing adsorption and oxidation kinetics. However, higher temperatures can also destabilize the flocs formed during coagulation. For example, at 60 °C, floc stability can decrease by 15–20% compared to ambient temperature conditions. The presence of dispersants, surfactants, and chelants alters the connections between the dye molecules and the solid phase. This can significantly influence treatment efficiency, with studies showing that the addition of 0.1–0.5% surfactants can improve the coagulation efficiency by 20–25% in jar-test experiments [36]. These physicochemical characteristics fundamentally determine which treatment technologies are effective for specific dye classes.

2.3 *Typical Composition and Characteristics of Dye-Laden Effluents*

Textile and dyehouse effluents exhibit intense color, high salinity, and complex organic loads that vary across mills and regions. Typical parameters include pH 6–11, COD 150–30,000 mg/L, BOD 80–6000 mg/L, TSS up to several mg/L, and TDS often exceeding 2–10 mg/L, with salts from dyeing operations reaching tens of grams per liter. Effluents are usually warm and may contain trace metals from dyes or equipment [21]. In reactive dyeing, producing 1 kg of cotton typically needs 70–150 L of water, 0.6–0.8 kg NaCl, and 30–60 g of dye, of which 10–40% remains unfixed, contributing to high conductivity and persistent color [36]. Effluent composition also fluctuates with fiber type, shade, and auxiliaries, demanding treatment systems resilient to variable loads [22].

2.4 *Toxicity and Biodegradability Concerns*

Azo dyes are the most broadly used class of synthetic dyes in the textile industry, representing more than 60% of the global dyes [41, 42]. Lack of dyeing routes leads to 15–50% of these dyes being released into wastewater without being fixed to the fibers [43]. When discharged untreated, azo dye effluents cause severe ecotoxicological effects, including reduced sunlight penetration in aquatic ecosystems, which impairs photosynthetic activity in algae and submerged plants [44]. Under anaerobic environmental conditions, azo bonds (–N=N–) are reductively cleaved by microorganisms to form aromatic amines, many of which are mutagenic and carcinogenic. For instance, 4-aminoazobenzene and related metabolites exhibit hepatocarcinogenicity in experimental animals [43]. Azo dyes can be ingested or absorbed via the skin by humans, resulting in the microbial biotransformation of dyes into toxic intermediates within the gut [45, 46]. Furthermore, bacteria on human skin have been shown to metabolize certain azo dyes into carcinogenic aromatic amines [47]. Aquatic life and higher trophic levels, including humans, may be impacted by the bioaccumulation of these harmful compounds in aquatic creatures, which can spread across the food chain [48]. Thus, azo dyes and their degradation products present a considerable environmental and public health risk, emphasizing the urgent need for integrated wastewater treatment systems capable of removing color, toxicity, and recalcitrant organics simultaneously [49]. The biodegradability of dyes varies significantly depending on their chemical class and molecular structure; therefore, decolorization cannot be regarded as a reliable indicator of mineralization or detoxification. Azo dyes often undergo rapid anaerobic reductive cleavage, forming aromatic amines that are frequently toxic and mutagenic. These intermediates require subsequent aerobic oxidation for complete degradation, which explains the common use of sequential anaerobic–aerobic treatment systems. In the absence of aerobic polishing, color removal may occur while COD and toxicity persist [50].

Disperse dyes, being nonionic and hydrophobic, exhibit low aqueous solubility and strong sorption to solids, which limit their bioavailability and hinder microbial degradation. Consequently, physico-chemical methods such as coagulation or adsorption often achieve faster color removal compared to COD reduction [51]. Metal-complex dyes demonstrate even better resistance to biodegradation due to the occurrence of coordinated metal ions (e.g., Cr, Co, Cu), which inhibit enzymatic attack and may introduce additional metal toxicity. As a result, these dyes typically require oxidative or adsorptive pretreatment before biological processing [52].

2.5 *Analytical Techniques for Dye Detection and Quantification*

Synthetic dye analysis, whether in solution or within fibers and fabrics, serves not only quality promise and direct regulation of the textile sector, but also underpins forensic science, environmental monitoring, food science, toxicology, and other contemporary uses. Synthetic dyes are used in foods, cosmetics, and pharmaceuticals, and they function as indicators in chemical processes. Notably, recent advances include their roles in photodynamic therapy, imaging, and dye-sensitized solar cells (DSSCs) [53, 54]. There are a few analytical methods currently used for the analysis of dyes, including current dye analysis, which relies on several techniques, including Fourier-transform infrared spectroscopy (FT-IR), ultraviolet–visible spectroscopy (UV–Vis), and chromatographic methods such as thin-layer chromatography (TLC) and high-performance liquid chromatography (HPLC) [55–60]. HPLC is among the most widely used techniques for dye analysis, particularly when coupled with mass spectrometry (HPLC–MS). Appropriate sample-preparation steps are also critical. For an in-depth treatment of HPLC–MS methods for dyes [58], this study highlights current developments in the field and focuses on the use of mass spectrometry (MS) for this purpose. MS uses a multiple of ionization methods, dector and mass to calculate the mass-to-charge ratio (m/z) of ionized materials. Because of their intended water solubility and structural resonance, which promotes stable charge creation, synthetic dyes are generally highly ionizable. Forensic, environmental, and pharmaceutical examinations require the sensitivity and specificity that MS offers. Additionally, its capability to perform in-situ measurements via direct ionization makes it especially advantageous. It can also detect and identify trace amounts of material even within challenging matrices like textiles, water, and soil [61, 62].

3 Effects of Dye-Contaminated Wastewater on the Environment and Human Health

A graphical overview of these interconnected impacts is presented in Fig. 2, illustrating how dye-contaminated wastewater acts as a major environmental and public-health threat. Dyes lower light penetration, reduce dissolved oxygen, and interfere with aquatic creatures' survival, reproduction, and variety after they are discharged into rivers and lakes. Many dyes and their toxic breakdown products accumulate in fish, invertebrates, and sediments, eventually biomagnifying through the food chain. People who are exposed to polluted water, food, or occupational settings face risks including skin irritation, respiratory problems, neurological effects, liver and blood toxicity, and increased cancer potential. Over time, these health and ecological damages translate into socioeconomic burdens for nearby communities, affecting livelihoods, healthcare costs, agricultural productivity, and overall quality of life.

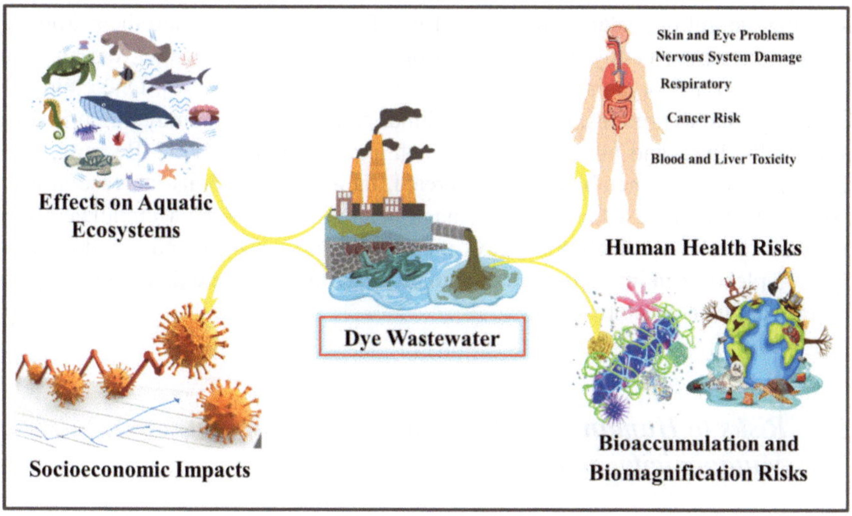

Fig. 2 Impacts of industrial wastewater pollution on humans, aquatic life, and the environment

3.1 *Impacts on Aquatic Environments*

The discharge of dye-contaminated effluents into aquatic systems significantly alters their physicochemical and biological balance. Textile dyes, particularly those that are non-biodegradable and highly soluble, reduce light penetration due to intense coloration, which limits the photosynthetic activity of aquatic plants and phytoplankton [63]. This reduction in photosynthesis consequently decreases the dissolved oxygen (DO) levels in the water, leading to oxygen depletion [64]. Low DO concentrations disrupt aerobic microbial activity and accelerate anaerobic decomposition, further deteriorating water quality. Moreover, the presence of organic pollutants and dye-derived aromatic amines elevates biochemical oxygen demand (BOD) and COD [65], intensifying hypoxic conditions that threaten aquatic life. Prolonged exposure to such conditions causes habitat loss, reproductive failure, and species migration, ultimately reducing biodiversity within aquatic ecosystems [66]. The toxic properties of many dyes, especially azo compounds, lead to bioaccumulation in aquatic organisms and biomagnification in the food chain, increasing their ecological impact [67, 68].

3.2 *Bioaccumulation and Biomagnification Risks*

Dye-contaminated effluents pose substantial bioaccumulation and biomagnification hazards within aquatic food webs. Many synthetic dyes, particularly azo, anthraquinone, and metal-complex varieties, are xenobiotic and highly persistent, resisting natural degradation processes [69]. These compounds and their aromatic amine derivatives readily adsorb onto sediments and organic matter, entering the tissues of benthic organisms through ingestion or absorption [70]. Once internalized, they bind to proteins, lipids, and cellular components, accumulating primarily in the gills, liver, and muscle tissues of fish and invertebrates [71]. Because these dyes are not easily metabolized or excreted, their concentrations increase at each trophic level, a process known as biomagnification [72]. The bioaccumulated dyes and their metabolites, including heavy metals such as Cr, Cu, and Ni found in metal dye complexes, can interfere with enzymatic activities, induce oxidative stress, and damage cellular structures [73].

3.3 Risks to Human Health: Endocrine Disruption, Mutagenicity, and Carcinogenicity

Textile dyes, especially azo dyes and benzidine-related dye classes, pose documented human-health hazards via metabolic activation to aromatic amines, direct genotoxicity, and endocrine activity. Azo bonds can be reductively cleaved by human gut microbiota and bacterial azoreductases, yielding aromatic amines with mutagenic and, in some cases, carcinogenic properties [45]. Evidence syntheses and toxicological reviews link such metabolites (e.g., benzidine, o-tolidine, o-dianisidine) to bladder and other cancers, supported by mechanistic data and occupational studies; authoritative evaluations classify benzidine as carcinogenic to humans and highlight risks from dyes metabolized to benzidine [74].

Beyond carcinogenicity, numerous azo dyes and dye mixtures exhibit mutagenicity in vitro and in vivo (e.g., chromosomal aberrations, DNA adduct formation), consistent with activation to electrophilic amines and oxidative-stress pathways [75]. Emerging evidence also indicates endocrine-disrupting activity for several commercially used colorants (anti-estrogenic/estrogenic signaling in receptor assays; reproductive endpoints in exposed workers), underscoring risk at low-dose, chronic exposures typical of environmental and occupational settings [76]. Regulatory and public-health listings (U.S. NTP Report on Carcinogens) flag benzidine and dyes metabolized to benzidine as cancer hazards, reinforcing the need for source control, substitution, and rigorous effluent treatment to reduce human exposure [77].

3.4 Socioeconomic Impacts and Public Health Concerns

Dye-contaminated effluents impose direct public-health burdens and indirect socioeconomic costs on communities living near textile clusters. Communities face elevated exposure to toxic dyes, heavy metals, and persistent organics, which are linked to dermatological, respiratory, and systemic diseases; recent reviews and case studies document increased health risks in river-adjacent settlements and food chains irrigated with contaminated water. These health burdens translate into lost productivity, medical expenses, and long-term disability, disproportionately affecting low-income workers and their families [78]. At the systems level, unchecked effluents degrade surface and groundwater, forcing municipalities and industries to invest in costly treatment, drill deeper wells, or purchase alternative supplies, raising water tariffs and production costs. Governance assessments in China and Bangladesh show that textile pollution intensifies water scarcity, conflict risks, and drinking-water safety concerns, creating fiscal pressure on utilities and eroding public trust [79].

Regionally, polluted rivers impair fisheries and agriculture, reducing yields and household income; clusters like Tirupur (India) illustrate how years of discharge contaminated aquifers and spurred expensive transitions (e.g., zero-liquid-discharge and centralized ETPs), with significant capital and operating costs ultimately borne by firms, workers, and consumers. Even after technology upgrades, legacy contamination and compliance gaps can sustain community-level health complaints and litigation [78]. Recent field evidence underscores persistent toxic exposures in textile hubs, e.g., heavy-metal-laden waters in South Asia associated with non-cancer and cancer risks, and PFAS detections well above advisory limits near garment factories, highlighting the need for stricter oversight, transparent monitoring, and health surveillance [36]. For example, pollution-driven shutdowns of dyeing units in Tirupur (India) resulted in industry losses exceeding USD 1 billion over a decade.

4 Conventional Treatment Technologies: Principles and Limitations

Conventional wastewater treatment technologies as shown in Fig. 3, the techniques are divided into chemical, physical, and biological processes along with hybrid systems. Adsorption, membrane filtration, coagulation-flocculation, and precipitation are examples of physical techniques that eliminate suspended particles, with sedimentation filtration ensuring clarified water. Chemical methods, including electrochemical oxidation, photocatalytic degradation, and chemical precipitation, use reactions and catalysts to convert pollutants into less harmful forms. Biological methods like aerobic and anaerobic degradation, enzymatic processes, and biofilm or membrane bioreactors employ microorganisms for the breakdown of organic matter. Hybrid and integrated systems combine these approaches to improve efficiency and the removal of pollutants.

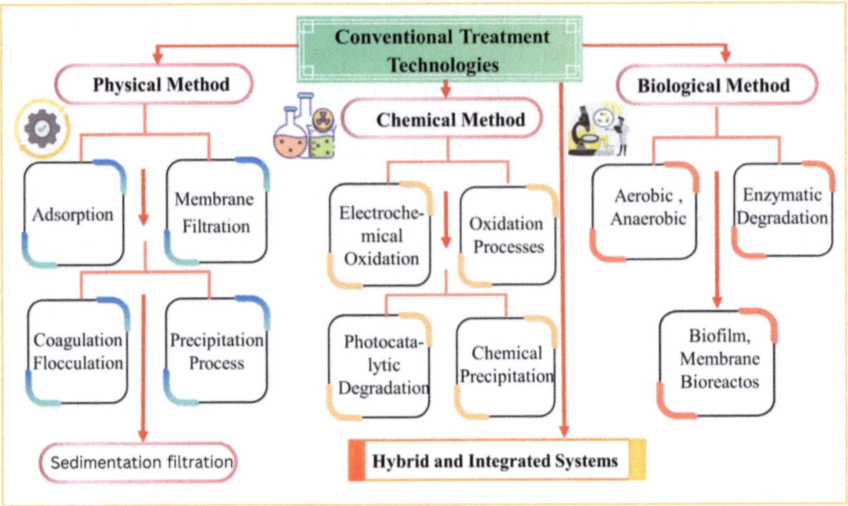

Fig. 3 Classification of conventional treatment technologies for wastewater and pollutant removal

4.1 Physical Treatment Methods

4.1.1 Adsorption

One very efficient way to clean industrial wastewater is by adsorption. Charged ions or molecules are drawn to and held on the solid surface of an adsorbent in this surface-based process. The dye molecules' adsorption is governed by various forces, including hydrogen bonding, electrostatic attractions, and van der Waals forces [80]. The adsorption efficiency mainly relied on the adsorbent properties, which typically possess a porous structure essential for the systematic and fast adsorption of dye molecules from wastewater [81, 82]. Commonly utilized adsorbents, such as silica gel, alumina, zeolite, and activated carbon, are highly effective in eliminating toxic dyes from contaminated water [83]. In addition to adsorption, other mechanisms, including complexation and ion exchange, are also employed for dye remediation [84–86]. It has been reported that Cu-doped Co_3O_4 nanoparticles (NPs) exhibit remarkable efficiency in the degradation of methylene blue (MB) dye. The incorporation of Cu ions into the Co_3O_4 lattice enhances the electronic structure and surface characteristics of the nanoparticles, thereby significantly improving their photocatalytic performance for dye-contaminated wastewater treatment [87]. In their study, Briao et al. reported that ZSM-5 zeolite serves as an effective adsorbent for the treatment of dye-containing wastewater. The material demonstrated notable removal efficiencies for various dyes, achieving degradation percentages of 81.2% for basic fuchsin [88], 75.3% for crystal violet, and 86.6% for MB, respectively. Several studies have demonstrated how different organic adsorbents and their composites can be applied to the degradation of aqueous solution dyes. Among these, conducting polymers in forms such as powders, aerogels, and biopolymers have been identified as especially promising materials due to their excellent adsorption capacity and structural flexibility [89–91].

4.1.2 Membrane Filtration

Membrane filtration is one of the most crucial physical techniques for removing dyes from textile wastewater [92, 93]. The method is based on the size exclusion principle, which states that molecules bigger than the pores in the membrane are kept on the membrane's surface. Microfiltration, a membrane-based separation technique, is particularly effective in the size range of 0.1–0.10 μm for removing dyes [94]. In this approach, microporous membranes facilitate the remediation of dyes and other contaminants from liquid effluents. Ultrafiltration, another membrane-based technology, has also demonstrated notable efficiency in dye removal. It is reported that ultrafiltration effectively reduces dye coloration in wastewater. Specifically, reactive black dye solutions treated with ultrafiltration ceramic membranes showed significant decolorization across various dye concentrations [95]. Similarly, reverse osmosis, a pressure-driven membrane filtration method, has been widely applied to the treatment of dye-polluted industrial effluent [96, 97]. Several studies have also affirmed the efficiency of various membrane filtration techniques in purifying dye-laden wastewater [98–100].

4.1.3 Coagulation-Flocculation (CF)

For many years, the CF process has been considered one of the most popular processes for handling textile wastewater, often serving as both a primary and pre-treatment stage in wastewater management systems. This technique is especially useful for significantly lowering the total suspended solids (TSS) load and decolorizing textile effluents [101, 102]. However, its application often results in the generation of substantial amounts of sludge, primarily due to excessive coagulant dosing. The raw wastewater's physicochemical properties, such as pH, concentration, coagulant type, and temperature, as well as mixing intensity and duration, have a significant impact on coagulation-flocculation performance [102]. The inclusion of a flocculation step is generally recommended to enhance the overall efficiency of coagulation by helping the aggregation of smaller particles into larger flocs, thereby refining sedimentation and clarification [103]. The primary disadvantage of the coagulation-flocculation method, despite its demonstrated efficacy, is the production of substantial amounts of sludge, which presents serious management and disposal issues [102, 104]. Consequently, the optimization of coagulant dosage is crucial to minimize sludge formation while maintaining high treatment efficiency.

4.1.4 Sedimentation and Filtration

Sedimentation and filtration are widely used as primary physical treatment steps for dye-containing textile effluents. Sedimentation relies on gravity settling to remove suspended solids, fibers, and large flocs, thereby reducing turbidity and total suspended solids (TSS) before subsequent treatment stages [105]. Filtration, using sand beds or granular media, further removes fine particulates that escape sedimentation and improves clarity, often serving as a polishing step after coagulation–flocculation or biological treatment [106]. However, both processes mainly target suspended matter and have limited efficiency for dissolved dyes and low-molecular-weight organics, which are typically responsible for color and toxicity in textile wastewater. For this reason, sedimentation–filtration trains are commonly integrated with coagulation–flocculation, adsorption, or advanced filtration (microfiltration, ultrafiltration, nanofiltration, and reverse osmosis) to achieve regulatory discharge limits and enable water reuse [107]. In recent studies, combining coagulation–flocculation–sedimentation with membrane or biosorbent-based filtration has shown improved removal of color, TSS, and recalcitrant dyes, highlighting the role of sedimentation and filtration as essential but non-sufficient unit operations within a multi-barrier treatment scheme [107].

4.2 Chemical Treatment Methods

4.2.1 Oxidation (Ozonation, Fenton, Electrochemical Oxidation)

Ozonation is a powerful single-oxidant process that rapidly decolorizes many dyes by cleaving chromophores; it is effective as a stand-alone or polishing step and can enable water reuse, but its mineralization of refractory organics is often incomplete unless combined with catalysts or peroxides [108]. Fenton (Fe^{2+}/H_2O_2) generates hydroxyl radicals ($^{\cdot}OH$) with high color/COD removal, especially at acidic pH; photo-Fenton and electro-Fenton improve Fe^{2+} regeneration and radical yield, reducing reaction time and energy per order [109]. Electrochemical oxidation (EO/EAOPs), including anodic oxidation on BDD or Ti-based anodes, electro-Fenton, and photoelectro-Fenton, achieves deep oxidation of diverse dyes and mixtures, with controllable kinetics and fewer chemicals, though electricity demand and electrode cost must be managed [110].

4.2.2 Advanced Oxidation Processes (AOPs)

In-situ preparation of very reactive radicals ($^{\cdot}OH$, $SO_4^{\cdot-}$) is the basis for AOPs, such as O_3/H_2O_2, UV/H_2O_2, EO/EF/PEF, sulfate-radical systems, sonocalysis, radiation, and photocatalysis, which non-selectively decrease dyes toward smaller acids and, under optimized conditions, CO_2 and H_2O. They offer high decolorization and COD removal across dye classes but require optimization of pH, oxidant dose, irradiation, and energy inputs [109]. The efficiency of AOPs in real textile effluents can be hindered by radical scavengers such as bicarbonate, chloride, and natural organic matter, which necessitates higher oxidant doses or the use of sulfate-radical-based systems that remain effective over a wider pH range. Hybrid AOPs (e.g., $O_3/Fe^{2+}/H_2O_2$, UV/EO) frequently outperform single processes in both efficiency and cost, with fewer sludge issues than coagulation-based methods [111]. Photocatalytic AOPs with modified or doped catalysts enhance visible-light utilization and mineralization efficiency. However, due to energy and chemical demands, AOPs are commonly applied as pre-treatment or polishing steps combined with biological or adsorption processes to improve sustainability.

4.2.3 Photocatalytic Degradation

The photocatalytic degradation efficacy for the semiconductor-solution interface depends on some factors, such as charge carrier dynamics, light absorption, and surface redox reactions. Photocatalysis, most commonly TiO_2 under UV/solar light, produces electron–hole pairs that form $^{\cdot}OH/^{\cdot}O^{2-}$ and mineralize a wide range of dyes; activity is enhanced by doping, heterojunctions, and supported catalysts to improve light use and charge separation [112]. Recent reviews report high decolorization/mineralization for azo and anthraquinone dyes and emphasize reactor design, light intensity, catalyst recycling, and toxicity of intermediates as critical scale-up factors [113].

4.2.4 Chemical Precipitation

Chemical precipitation (often operationalized as coagulation–flocculation–decantation using metal salts or polymers) efficiently removes insoluble dye aggregates, dye–metal complexes, auxiliaries, and especially dissolved metals via hydroxide/sulfide/carbonate formation; it is robust and simple but generates sludge requiring safe disposal [114]. For wastewaters with significant heavy-metal content (from metal-complex dyes or auxiliaries), precipitation remains a mature, first-line option and is frequently paired with AOPs or adsorption for residual color/COD polishing [115].

4.3 Biological Treatment Methods

Biological treatment methods are often seen as more effective compared to physical or chemical approaches. These methods are easier to apply, produce less sludge, require fewer chemicals, and are more cost-efficient, energy-saving, and environmentally sustainable. Furthermore, the byproducts from microbial metabolism are generally non-toxic [116]. Furthermore, biological processes can completely mineralize dyes and are economically viable in developing countries [117]. Breaking down refractory organic dyes into innocuous compounds is the primary goal of biological therapy. A significant advantage of this method is the use of microorganisms with strong biodegradation capabilities, either alone or in combination, to effectively treat wastewater contaminated by dyes [118].

4.3.1 Bacterial-Assisted Degradation of Dye-Containing Wastewater.

Several bacterial species have demonstrated greater efficiency in degrading dye-contaminated wastewater compared to other microorganisms [119]. The effectiveness of bacterial dye degradation depends on their adaptability and capacity to break down dyes under existing environmental conditions [120]. Bacteria like *Klebsiella*, *Pseudomonas*, and *Bacillus* efficiently biodegrade azo dyes due to easy cultivation, fast growth, and strong catalytic activity [116, 121]. The main benefits of employing bacteria for dye degradation include their varied enzymatic capabilities to mineralize azo dyes in wastewater, their simplicity of cultivation, and their quicker growth rate as compared to other microorganisms [122]. Table 2 demonstrates the effectiveness of various bacterial strains in degrading a wide range of synthetic dyes under optimized environmental conditions. Species such as *Bacillus*, *Pseudomonas*, *Klebsiella*, *Streptomyces*, and *Bacillus cereus* demonstrate high decolorization efficiencies, often exceeding 90% for reactive, disperse, and direct dyes. [123–125]. These microorganisms utilize enzymatic pathways such as azoreductases, laccases, and peroxidases to reduce dye molecules, with performance strongly influenced by pH, temperature, nutrient availability, and dye concentration. Biofilm-forming *Pseudomonas* strains show enhanced stability and removal efficiency, suggesting their potential for continuous-flow bioreactors [126–128]. Overall, the data underscore the promise of bacterial bioremediation as an environmentally friendly and economical alternative to chemical treatment methods for dye-contaminated wastewater.

4.3.2 Enzymatic Degradation

Enzymatic bioremediation has become a novel, effective, and environmentally sustainable method for treating dye-contaminated wastewater [129]. This method involves utilizing the metabolic capabilities of microorganisms or genetically modified strains to produce specific enzymes capable of degrading dyes. Through enzymatic catalysis, these dyes are transformed from toxic or hazardous compounds into harmless or non-toxic forms [129]. The enzymatic degradation of azo dyes has been widely studied, focusing on the characterization of various enzyme systems responsible for breaking down these complex molecules [129, 130]. The enzyme azoreductase was found by Singh [131] to be a crucial catalyst in the reductive cleavage of azo linkages (–N=N–) within dye complexes, which transforms aromatic amines into colorless, less toxic molecules. Similarly, the enzyme laccase has demonstrated

Table 2 Summary of bacteria-assisted dye degradation in wastewater

Bacterial species	Dye	Dye removal (%)	Key conditions	References
Bacillus albus DD1	Reactive black 5 (RB5)	98	38 h, pH 7, 40 °C, 50 mg/L RB5, 1% yeast extract	[123]
Pseudomonas aeruginosa	Methyl red	88.37 (optimized); 81.49 (initial screen)	~3 days, pH 9, 40 °C, 20 ppm MR, nutrients and redox mediator	[124]
Pseudomonas aeruginosa MT-2	Congo red	98.2	48 h, initial 50 mg/L, optimized conditions	[125]
Pseudomonas aeruginosa (biofilm)	Congo red	75.74	10 h, 50 ppm CR, preformed biofilm on glass	[121]
Klebsiella pneumoniae GM-04	Disperse blue 284	95	24 h, pH 7, 200 ppm	[127]
Streptomyces albidoflavus 3MGH	Reactive orange 122; direct blue 15; direct black 38	61.6; 63.7; 54.5	At 0.2 g/L dye; rates decline at higher concentrations	[128]
Bacillus cereus MT-4	Direct blue-1	97	Optimized culture; 5% v/v inoculum	[125]

great potential in reducing a wide range of toxic textile dyes [132]. Furthermore, peroxidase enzymes, specifically manganese peroxidase and lignin peroxidase, have been widely considered for their capacity to oxidatively degrade dyes [133].

4.3.3 Biofilm and Membrane Bioreactors

A biofilm is a self-produced matrix identified as extracellular polymeric substances (EPS) that traps an organized community of bacteria. This EPS matrix protects microbes from pollutants, predatory protozoa, and various environmental stresses on both biotic and abiotic surfaces. It primarily comprises proteins, lipids, polysaccharides, nucleic acids, humic compounds, and surfactants [134]. Water, constituting about 97% of the biofilm, facilitates nutrient transport within the matrix [135]. The EPS matrix is typically between 0.2 and 1.0 μm thick, but the biofilm layer is typically between 10 and 30 μm thick. EPS accounts for 65–95% of the total biofilm composition, with microorganisms forming the remainder. FTIR analysis reveals that polysaccharides and proteins are the dominant components of biofilms [135]. Biofilms are classified into bacterial, algal, fungal, and symbiotic types based on their dominant microbial constituents. Bacterial biofilms are the most prevalent, contributing significantly to bioremediation and wastewater treatment. Algal biofilms enhance nutrient removal and oxygen production, while fungal biofilms degrade complex organic compounds through strong enzymatic activity. Symbiotic biofilms, such as fungi–bacterial or protozoan–bacterial consortia, display synergistic metabolism that enhances pollutant removal efficiency. Together, these diverse biofilm systems form the foundation of many sustainable biotechnological applications. Biofilm-based bioreactors have appeared as an effective technology for wastewater treatment, offering improved pollutant removal, reduced sludge creation, smaller hydraulic retention times (HRT), and high microbial diversity supported by extracellular polymeric substances (EPS). The presence of dense active biomass and stable biofilm structures further improves system performance, making these reactors widely applicable for treating highly contaminated wastewater [136].

Biofilm-based bioreactors are categorized into several types, including membrane, moving-bed, fluidized-bed, and microbial fuel cell systems, as shown in Fig. 4. Membrane bioreactors (MBRs) utilize biofilms for effective separation and pollutant degradation and are further classified into oxygen- and hydrogen-supplying or novel MBRs for specific applications. Moving-bed and fluidized-bed reactors provide large surface areas and enhanced mass transfer, promoting stable biofilm growth and efficient treatment performance. Microbial fuel cells, meanwhile, harness biofilm activity to generate bioelectricity while simultaneously treating wastewater. Collectively, these bioreactor types exhibit the adaptability and efficiency of biofilm-based technologies in environmental remediation.

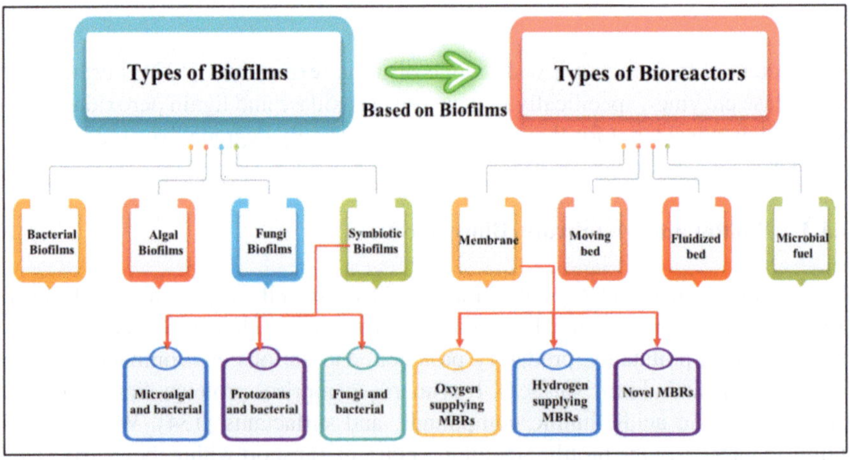

Fig. 4 Overview of various types of biofilms and corresponding bioreactors

4.4 Hybrid and Integrated Treatment Systems

To advance removal efficiency, hybrid wastewater treatment systems combine chemical, biological, and physical processes and meet strict discharge standards, which is the most suitable approach to treating dye wastewater. For example, ozonation followed by biological treatment increases the biodegradability of persistent organic pollutants, allowing the bioreactor to complete mineralization [137]. Anaerobic–aerobic sequences such as UASB followed by activated sludge or constructed wetlands recover biogas and remove bulk COD before final nutrient and pathogen polishing [138]. Biofilm-based systems (MBBR/IFAS) improve nitrification stability under fluctuating loads [139]. In decentralized applications, combining constructed wetlands with anaerobic or oxidative pretreatment offers reliable polishing with low operational cost. Electrocoagulation with biological polishing reduces color, metals, and toxicity while maintaining downstream treatment performance [137]. Although hybrid systems enhance pollutant removal, energy efficiency, and resilience, they require careful design to avoid excess chemical use, by-product formation, and operational complexity.

4.5 Comparative Analysis of Conventional Methods: Advantages, Limitations, and Case Studies

Conventional activated sludge (CAS) often fails to remove persistent dye color and can leave toxic aromatic amines. Hybrid systems overcome these limits by pairing ozonation with biological treatment for stronger decolorization and using anaerobic, aerobic sequencing to degrade amines and reduce toxicity, as shown in Table 3. MBBR/MBR configurations provide better stability under shock loads [140–142], while EC reduces dye/metal toxicity and supports EC + NF/RO reuse trains [143]. Pretreating wastewater before UF followed by NF/RO improves the quality of reusable water [144], and using anaerobic treatment together with optimized aeration helps lower the overall energy consumption [145]. Constructed wetlands provide a low-energy option for final polishing of wastewater [146], and in addition, controlling dye adsorption onto sludge using activated carbon or biochar together with electrocoagulation can reduce sludge-related disposal problems [147].

Table 3 Key challenges in dye wastewater treatment and corresponding hybrid solutions relative to conventional CAS systems

Challenges	Conventional CAS	Hybrid approaches	References
Strong color; stubborn azo dyes	Many dyes don't biodegrade in CAS, so color passes through	Ozonation + Biological (either O_3 → bio or bio → O_3); Activated Carbon as polish	[140]
Toxic dye break-products (aromatic amines)	In CAS, partial breakdown can leave toxic intermediates that hurt microbes	Anaerobic → Aerobic sequence (e.g., UASB/AnSBR then aerobic/MBR/MBBR)	[141]
Big swings in COD and color (shock loads)	CAS flocs lose performance during shocks; nitrification also wobbles	MBBR/IFAS retrofits (biofilm = tougher), or MBR; add EC/O_3	[142]
Metals + dye toxicity	Metals and high dye loads inhibit CAS bugs; flocs destabilize	Electrocoagulation (EC) before/after bio; EC + NF/RO trains	[141, 143]
High salts (TDS) and faint leftover color	CAS cannot remove dissolved salts; faint chromophores remain	UF → NF/RO after EC/O_3/AC pretreatment	[144]
Energy use (aeration is expensive)	Aeration is the largest energy load in CAS	Anaerobic pre-treat (UASB/AnSBR) + smarter aeration control/blowers	[145]
Low-resource polishing/rural reuse	CAS plants need power and skilled ops; not always available	Constructed wetlands as tertiary dye-polishing; phytoremediation cells	[146]
Dye ending up in sludge	Dyes can adsorb to sludge; disposal becomes a color problem	Use AC/biochar and EC to shift removal into water line; stabilize/oxidize side-streams	[147]

Note UASB stands for upflow anaerobic sludge blanket reactor, AnSBR for anaerobic sequencing batch reactor, MBR for membrane bioreactor, MBBR for moving bed biofilm reactor, IFAS for integrated fixed-film activated sludge, RO for reverse osmosis, AC for activated carbon, EC for electrocoagulation, COD for chemical oxygen demand, and TDS for total dissolved solids

5 Emerging Nanomaterials and Advanced Sorbents in Dye Removal

5.1 Introduction to Nanotechnology in Wastewater Treatment

Nanotechnology has reshaped dye-wastewater treatment by enabling high-surface-area sorbents and photocatalysts that couple fast mass transfer with tunable surface chemistry for adsorption, advanced oxidation, and easy post-separation. Carbon nanomaterials (graphene graphene oxide) deliver rapid, high-capacity dye uptake and robust regeneration when functionalized or composited, while also serving as conductive scaffolds for hybrid catalysts [148]. Semiconductor photocatalysts, classically TiO_2 and ZnO, and newer systems like g-C_3N_4 drive visible-light dye degradation via reactive oxygen species, with performance boosted by doping, heterojunctions, and carbon coupling to overcome recombination and extend light absorption [149]. Two-dimensional MXenes (e.g., $Ti_3C_2T_x$, Nb_2C_{tx}) and related 2D hybrids offer exceptional charge transport and abundant surface terminations for synergistic adsorption–photocatalysis, emerging as versatile platforms for dye capture and degradation [150]. Porous crystalline adsorbents, MOFs and COFs, provide designer pore architectures and functional sites that enable selective dye adsorption and photocatalytic pathways; recent work emphasizes greener syntheses and stability under realistic waters [151]. Magnetic nanomaterials (e.g., Fe_3O_4 and ferrites) add a practical advantage: rapid magnetic recovery and reuse of sorbents and catalysts from complex matrices [152]. Collectively, these nanostructures enable hybrid processes (adsorption–photocatalysis, Fenton-like and photo-Fenton, and membrane nanocomposites) that achieve high decolorization and mineralization at lower doses and shorter contact times; key challenges now center on stability, anti-fouling, leaching/toxicity, and scalable, low-waste synthesis [113]. To address these challenges, metal and metal oxide nanoparticles (NPs), carbon-based nanomaterials, metal–organic frameworks (MOFs), and nanocomposite hybrid systems are increasingly being explored.

5.2 *Metal and Metal Oxide Nanoparticles*

Metal/oxide NPs remove dyes via two main routes: (i) adsorption on high-area, surface-functional sites and (ii) advanced oxidation (photocatalysis or Fenton-like chemistry) that mineralizes dyes to CO_2, H_2O, and small ions. TiO_2 and ZnO dominate photocatalysis because of suitable band structures and stability; visible-light activity is often achieved by doping, heterojunctions, or carbon co-sensitization to suppress electron–hole recombination. The UV-active titanium dioxide remains a benchmark reference for photocatalysis and dye degradation; visible-light performance is routinely boosted by building heterojunctions with graphitic carbon nitride (g-C_3N_4) and by heteroatom doping (e.g., Ag, Fe, P) to narrow the effective band gap and improve charge separation. MOF-TiO_2 hybrids further enhance interfacial charge transfer and reusability under solar/visible light [113]. ZnO often matches or exceeds 2 for many azo and cationic dyes but is prone to photocorrosion under illumination; current remedies include defect/morphology engineering, noble-metal or carbon-dot coupling, and Z-type-II heterojunctions with ferrites and sulfides. Representative reviews and composites (e.g., $NiFe_2O_4$/ZnO) show improved quantum efficiency and durability under visible light [153]. Magnetite (Fe_3O_4) serves both as an adsorbent and a heterogeneous Fenton/Fenton-like catalyst; its magnetic separability enables practical recovery. It is frequently integrated into photocatalytic heterostructures (e.g., Fe_3O_4/TiO_2, Fe_3O_4/ZnO, or Fe_3O_4@g-C_3N_4), delivering combined adsorption–photo-Fenton or adsorption–photocatalysis with straightforward magnetic recycling [154]. CuO and other narrow-gap oxides broaden visible-light response and, when coupled with TiO_2 or ZnO, form efficient interfacial charge-transfer junctions that accelerate dye mineralization; numerous reports now use green/biogenic synthesis routes for such CuO–TiO_2 systems [155].

5.3 *Carbon-Based Nanomaterials (Graphene Oxide, CNTs, CQDs)*

The many oxygenated functional groups (–COOH, –OH, and epoxide) found in graphene oxide (GO) allow for electrostatic attraction, hydrogen bonding, electrostatic attraction, and π–π interactions with organic pigments and heavy metals. Its hydrophilicity promotes dispersion in water and fast binding. Composite forms (e.g., GO-biopolymers, GO-clays, magnetic GO) improve adsorption capacity and allow easy magnetic separation and regeneration [156]. Dyes mainly adsorb by hydrophobic interaction and π–π stacking by carbon nanotubes (CNTs). Oxidation or amine/polymer functionalization significantly enhances surface polarity, adsorption selectivity (cationic versus anionic dyes), and kinetics. Packed-bed and fixed-bed column studies confirm their potential for industrial-scale wastewater treatment [157]. Carbon Quantum Dots (CQDs) provide surface functional groups and excellent electron-transfer properties, enabling dual roles as adsorbents and visible-light photosensitizers. CQD/metal-oxide heterostructures (e.g., CQD-TiO_2, CQD-ZnO) significantly enhance dye degradation efficiency and improve photostability and reusability under visible light [158].

5.4 *Metal–Organic Framework (MOFs) and Covalent Organic Frameworks (COFs)*

The crystalline, ultraporous MOFs show high dye uptakes through size/shape matching and functional groups that enable electrostatic and π–π interactions; MOFs (including bimetallic MOFs) often outperform conventional adsorbents for dye-laden effluents and are promising as membranes/films and electrospun composites for polishing steps. Many MOFs also act as photocatalysts via linker-to-metal charge transfer or as precursors to porous carbons that add robustness and recyclability [159]. The COFs provide designer, rigid pores with extended π-conjugation and tunable surface charge; negatively charged or π-rich channels favor adsorption of cationic/aromatic dyes, while the conjugated backbones enable visible-light photocatalysis. Recent reviews detail 2D/3D COFs, scalable syntheses, and mechanism-driven enhancements (charge transport, radical pathways), with experiments and simulations guiding linker/topology choices; metal-free COF photocatalysts and PMS-activated systems demonstrate efficient dye degradation and good stability [160].

5.5 Clay-Based and Silica-Based Nanomaterials

Because of their layered architectures and large surface areas, naturally occurring clays such as montmorillonite, bentonite, kaolinite, and halloysite nanotubes (HNTs) are often utilized in color removal and cation-exchange capacities, which facilitate intercalation and ion-exchange adsorption. Halloysite nanotubes, in particular, provide hollow tubular channels that enhance the uptake of bulky dye molecules. Recent studies show that magnetic clay hybrids (e.g., Fe_3O_4-HNTs) and clay–biopolymer composites (e.g., chitosan–bentonite) improve adsorption capacity, mechanical stability, and enable magnetic or mechanical recovery and reuse across broad pH ranges [161]. The silica mesoporous materials, especially MCM-41 and SBA-15, exhibit uniform, tunable pore channels and high accessible surface areas, enabling rapid adsorption of organic dyes. Surface functionalization (e.g., amine, thiol, or polycarboxyl modification) tailors the material to selectively remove anionic or cationic dyes through electrostatic attraction and hydrogen bonding. These functionalized silicas typically show fast adsorption kinetics, high capacity, and good regeneration stability over multiple cycles [162].

5.6 Magnetic Nanomaterials for Adsorptive and Catalytic Removal

Magnetic nanomaterials, with high adsorption effectiveness in particular, Fe_3O_4 and spinel ferrites (MFe_2O_4), have garnered a lot of interest for dye-contaminated wastewater treatment and, importantly, their ability to be rapidly separated and reused through magnetic recovery. These materials function both as adsorbents, for example, Fe_3O_4 supported on activated carbon, chitosan, or other biopolymers [163]. Magnetically recyclable photocatalysts, when coupled with semiconductor systems such as ZnO, TiO_2, and g-C_3N_4. Such heterostructures promote efficient charge separation, enabling enhanced photocatalytic degradation of dyes while maintaining structural stability over repeated cycles [164]. Recent reviews highlight their advantages in reusability, stability under varying pH, and hybrid advanced oxidation performance, positioning them as strong candidates for scalable wastewater remediation [165]. Additionally, green synthesis approaches using plant extracts and bio-templates have been reported, offering reduced environmental impact and improved compatibility with circular-economy manufacturing [166]. Overall, magnetic nanomaterials combine high removal efficiencies with simplified post-treatment handling, making them promising for practical, sustainable dye effluent management.

5.7 *Functionalized and Doped Nanomaterials*

5.7.1 CQD and W-doped and TNTs

Carbon quantum dots (CQDs, 2–10 nm) possess surface COOH/–OH/–NH_2 groups that provide dye adsorption sites and strong interaction with semiconductors. Their broad visible-light absorption and fast electron-transfer capacity enable photosensitization of TiO_2, promoting electron injection and reducing e^-/h^+ recombination, which increases ROS generation and accelerates dye photodegradation, such as methyl orange and MB, compared to pristine TiO_2 [167, 168]. The W^{6+}-related defect states are created by tungsten doping in TiO_2 nanotubes, extending light absorption into the visible range and suppressing charge recombination, resulting in higher photocatalytic activity than undoped TNTs [168]. When W-TNTs are integrated with conductive supports (e.g., reduced graphene oxide), interfacial electron transport is further enhanced, yielding faster dye degradation [169].

5.7.2 Cu-Doped Co_3O_4 and Other Hybrid Catalysts

Spinel Co_3O_4 is a visible-light-active p-type photocatalyst; however, Cu doping (Cu^{2+}/Cu^+) effectively modifies the Co^{2+}/Co^{3+} ratio, introduces oxygen vacancies, and increases visible-light absorption and ROS generation. $Cu_XCo_{3-X}O_4$ materials consistently show higher MB degradation efficiencies and faster reaction kinetics than undoped Co_3O_4 under visible irradiation [170]. The 15% Cu-doped Co_3O_4 sample exhibited the highest photocatalytic performance, obtaining about 89% MB degradation after 80 min when exposed to visible light, with efficiency improving progressively with Cu concentration [87].

5.8 *Nanocomposites and Hybrid Systems for Synergistic Adsorption and Photocatalysis*

Nanocomposites that integrate high surface-area adsorbents (e.g., biochar, activated carbon, clays, MOFs, polymers) with photocatalysts (such as TiO_2, g-C_3N_4, and spinel oxides) operate via an "adsorb then photodegrade" mechanism, where the adsorbent first concentrates dye molecules near the catalyst surface, increasing local reaction rates and enhancing light-driven degradation efficiency [171, 172]. Conductive carbon scaffolds (e.g., rGO, CNTs) additionally promote rapid electron transport and reduce electron–hole recombination, resulting in higher degradation efficiencies, improved pH tolerance, and better stability in real-water conditions compared to single-component materials [173].

6 Mechanisms of Dye Removal Using Nanomaterials

6.1 Adsorption Mechanisms: Electrostatic Interactions

Electrostatic interactions govern dye adsorption because both dye molecules and nanomaterial surfaces exhibit pH-dependent charges. CQDs, GO, and functionalized CNTs develop a negative surface charge at higher pH, enabling strong attraction of cationic dyes such as MB, while protonated surfaces favor anionic dye adsorption. Similarly, TNTs, W-doped TiO_2, and Fe_3O_4-based magnetic composites adjust their surface charge around the isoelectric point, allowing selective dye binding and easy separation [167, 174–176]. Therefore, controlling pH and surface functional groups is essential for maximizing electrostatic dye removal efficiency.

6.2 Photocatalytic Degradation Pathways

Under light, nanomaterials generate electron–hole pairs as shown in Fig. 5 that form ROS ($^{\cdot}OH$, $^{\cdot}O_2^-$), which oxidatively break dye chromophores into CO_2 and H_2O [177]. Carbon-based nanomaterials (GO, CNTs, CQDs) enhance dye adsorption and accelerate charge transfer, suppressing e^-/h^+ recombination. CQDs and W-doping extend visible-light activity, while TiO_2 nanotubes improve charge transport [167, 178]. Magnetic nanomaterials (e.g., $Fe_3O_4@TiO_2$) also allow simple catalyst recovery and reuse [167].

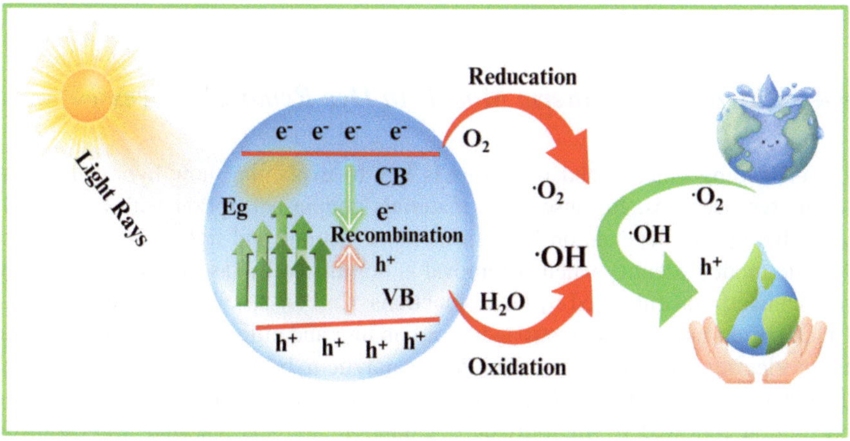

Fig. 5 Schematic illustration of photocatalytic dye degradation

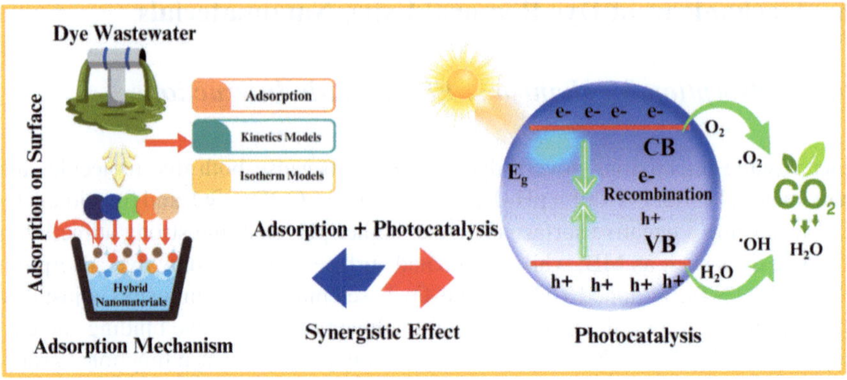

Fig. 6 Schematic illustration of the synergistic effect in hybrid nanomaterials for dye wastewater treatment

6.3 Synergistic Effects in Hybrid Nanomaterials

In dye wastewater treatment, hybrid nanomaterials combine adsorption and photocatalysis, as illustrated in Fig. 6, to improve pollutant removal efficiency. Carbon-based nanostructures (GO, CNTs, CQDs) form conductive heterojunctions with oxides that enhance dye adsorption (π–π interactions) and accelerate charge transfer, reducing e^-/h^+ recombination [179, 180]. CQDs further act as visible-light sensitizers and electron shuttles, while W-doping and TiO_2 nanotubes (TNTs) improve bandgap utilization and 1D charge transport [181]. Magnetic hybrids (e.g., $Fe_3O_4@TiO_2$) additionally allow simple catalyst recovery and stable reuse [182].

6.4 Kinetic and Isotherm Models in Dye Removal Processes

Understanding kinetics and isotherms is key to explaining how hybrid nanomaterials remove dyes. Because these materials couple adsorption with photocatalysis, dye uptake depends on both surface interactions and light-driven reactions. Kinetic models identify whether removal is dominated by fast physical adsorption or chemisorption enhanced by heterojunctions. Carbon-based hybrids (GO, CNTs, CQDs) often fit PSO due to stronger π–π bonding and efficient charge transfer that limits electron–hole recombination. Photocatalytic hybrids such as W-doped TiO_2, TiO_2 nanotubes, and $Fe_3O_4@TiO_2$ typically show two-stage behavior: rapid surface adsorption followed by light-driven degradation. Intraparticle and film-diffusion models help determine whether diffusion through nanotube channels, doped structures, or surface reactions controls the rate. Isotherm models explain equilibrium dye uptake. Langmuir behavior is common in TiO_2-based hybrids with uniform active sites and monolayer adsorption, while Freundlich fits carbon-rich hybrids with heterogeneous surfaces that support multilayer adsorption before photocatalytic breakdown. Overall, kinetics and isotherms clarify how hybrid structures, carbon matrices, dopants, nanotubes, and magnetic cores govern adsorption strength, charge transport, and reaction pathways, ultimately improving dye removal efficiency. These mechanisms collectively explain why hybrid nanomaterials outperform single-function systems in both adsorption capacity and degradation kinetics.

7 Techno-Economic and Environmental Considerations

7.1 Cost–benefit Analysis of Conventional vs. Nanomaterial-Based Processes

Recent advancements in nanotechnology have significantly improved wastewater treatment, with nanomaterials (e.g., nanoparticles, nanocomposites, and nano-enabled membranes) achieving up to 99% contaminant removal, outperforming conventional methods (60–80% efficiency) [183]. However, a significant disadvantage is still the high cost of producing and functionalizing nanomaterials, with synthesis costs ranging from USD 20–200 per kg, driving up the overall treatment cost per cubic meter of water [184]. To make nanomaterials economically viable, reuse and recycling are crucial. Reusing nanomaterials for multiple cycles can cut per-cycle costs by more than 70%, making them competitive, particularly in small-scale applications or as a polishing step in conventional systems [183]. Hybrid systems that integrate conventional treatments with nanomaterials can reduce both costs and environmental impact. By using nanomaterials only as a polishing/enhancement step after conventional methods (e.g., coagulation, activated carbon), these systems minimize nanomaterial usage while still benefiting from their high efficiency. This significantly reduces costs and the environmental risk of using large quantities of nanomaterials. Moreover, hybrid systems facilitate more efficient recycling and reusing of nanomaterials, enhancing both their cost-effectiveness and reducing environmental disposal risks [185].

In conclusion, while nanomaterial-based systems offer high performance, their high costs and environmental concerns make hybrid systems the most practical solution. Combining conventional methods with nanomaterials allows for a more economically and ecologically viable treatment process, optimizing benefits while minimizing risks [186].

7.2 Energy Demand, Operational Complexity, and Scalability

Conventional dye-treatment systems consume significant energy, mainly through aeration, pumping, mixing, and electrochemical operations, which become more intensive under high pollutant loads or strict discharge standards [187]. Biological MBR/MBBR systems save space but require higher energy and operational control due to membrane fouling and biofilm management, while conventional membranes demand careful pre-treatment and regular maintenance [188]. Nanomaterial-based processes can lower energy use when driven by sunlight or rapid adsorption, and nano-membranes often provide higher flux with reduced fouling [173]. However, they require tighter operational control to prevent nanoparticle release and maintain surface stability [183]. Large-scale application remains limited because nano-processes are mostly proven at lab or pilot levels, though hybrid systems such as EC combined with nano-membranes or photocatalytic polishing offer a practical and scalable way forward [189].

7.3 Life Cycle Assessment (LCA) and Environmental Footprint

LCA is increasingly employed to assess dye wastewater treatment technologies by accounting for energy use, chemical production, infrastructure needs, and sludge management factors not captured by removal efficiency alone. Studies on Fenton and electro-Fenton processes show that electricity and hydrogen peroxide production are major environmental hotspots despite strong treatment performance [190]. Integrated systems, such as jet-loop reactors with ozonation and membrane units, can lower overall impacts, but only when chemical use and the electricity mix are optimized. LCA frameworks generally favor options that generate less sludge and enable higher water reuse, even if they have slightly higher costs [191, 192]. For nanomaterial-based methods, limited LCA data suggest that material synthesis, potential ecotoxicity, and end-of-life handling may offset some performance benefits [183]. Future assessments must include nanoparticle fate, regeneration frequency, and release risks to accurately determine their environmental footprint.

7.4 Resource Recovery and Valorization of Dye-Laden

Resource recovery is increasingly recognized as a core pillar of CE-oriented dye wastewater management, where waste streams are reframed as secondary resources rather than disposal liabilities. Conventional coagulation and adsorption processes generate dye-rich sludge, which can be regenerated through thermal, chemical, biological, or electrochemical routes; however, these methods often require additional energy and reagents, partially offsetting their sustainability benefits [193]. From a CE perspective, greater value retention can be achieved by recovering dyes, metals, or functional chemicals from regenerant streams, enabling their reuse in upstream industrial processes and closing material loops [194]. Beyond regeneration, increasing emphasis is placed on the valorization of dye-laden residues into functional materials, particularly the conversion of sludge into biochar or carbon-based adsorbents. Such sludge-derived biochars not only divert waste from landfilling but also serve as low-cost sorbents for dyes, heavy metals, and emerging contaminants, effectively extending the service life of the original material across multiple cycles of use [195]. Spent nano-adsorbents and catalysts can also be regenerated or incorporated into new composites, reducing the need for fresh materials. At the system scale, ZLD and membrane-integrated treatment schemes further strengthen CE implementation by simultaneously recovering high-quality water and concentrated salts suitable for industrial reuse. These closed-loop configurations minimize freshwater demand, reduce waste discharge, and improve economic viability by transforming treatment plants into resource recovery facilities rather than end-of-pipe solutions [196]. Collectively, such approaches align dye wastewater management with CE principles of waste minimization, material circulation, and value creation, supporting long-term environmental and economic sustainability.

8 Future Perspectives and Directions in Technical Textiles Wastewater Treatment

Advancing dye wastewater treatment, particularly from technical textiles, demands technologies that are efficient, scalable, and economically viable. As given in Fig. 7, future work focuses on multifunctional, recyclable nanomaterials and the use of artificial intelligence (AI) and machine learning (ML) for predictive modelling and process optimization. The addition of technical textiles into wastewater treatment processes should also be explored, particularly considering the specific challenges posed by the dyes used in textile manufacturing. Integrating resource-recovery units will strengthen circular economy (CE) goals by reclaiming salts, dyes, and reusable water. Green synthesis routes using low-toxicity, biomass-based precursors will further reduce environmental impact. Strong regulatory frameworks and standardized protocols are also needed to confirm the safe adoption of nano-enabled technologies in technical textile manufacturing. Ultimately, bridging laboratory research with industrial-scale testing will determine the successful translation of these innovations into practice.

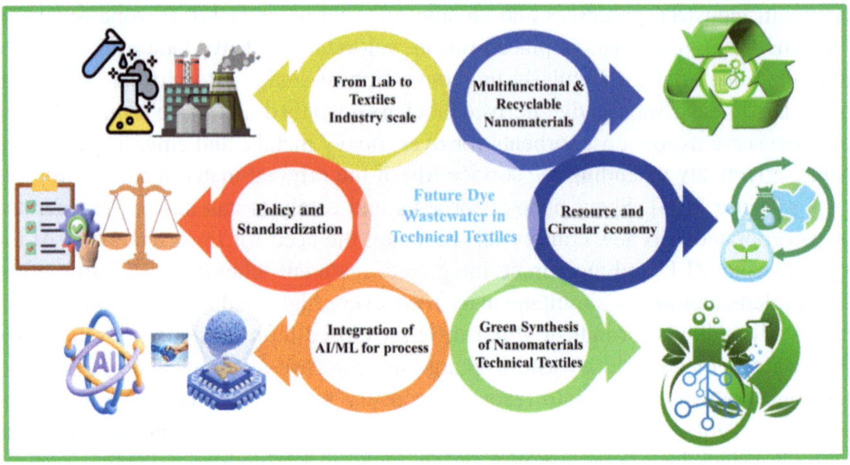

Fig. 7 A roadmap for the future of dye wastewater treatment: innovation, recycling, and sustainable nanotechnology

8.1 *Design of Multifunctional, Selective, and Recyclable Nanomaterials*

The development of nanomaterials highlights the future research that combines adsorption, photocatalysis, and antimicrobial action into a single platform, specifically targeting the dyes used in technical textiles. These materials should exhibit high selectivity for anionic or cationic dyes, while enabling easy magnetic or chemical regeneration. Recent reviews highlight metal oxides, MOFs, carbon-based nanomaterials, and their hybrids as promising multifunctional, recyclable adsorbents and photocatalysts for dye removal in textile wastewater [197]. Robust designs with minimal leaching and stable performance over many cycles are essential for scaling up these materials to meet real-world industrial needs in technical textiles.

8.2 *Integration of AI/ML for Process Optimization and Predictive Modelling*

The integration of machine learning (ML) with artificial intelligence (AI) is essential for modeling the complicated, non-linear behavior in wastewater treatment, particularly for technical textiles. AI/ML can be used to accurately predict dye removal efficiency, energy consumption, and sludge production, based on real-time operational data [198]. These tools can optimize key treatment parameters such as contact, pH value, dosages, and aeration, improving performance and cost-efficiency. AI and ML can also assist in fault detection, decision-making, and real-time process optimization, contributing to the more effective treatment of textile effluents in nano-enabled systems.

8.3 *Integrating Treatment with Resource Recovery and Circular Economy Principles*

The future management for dye wastewater in technical textiles will increasingly focus on integrating ZLD, salt and dye recovery, and water reuse within treatment systems. Emerging membrane-based ZLD systems, evaporators, and hydrate-based processes are capable of recovering high-purity salts and reusable dyes while meeting stringent discharge limits [199]. Embedding resource-recovery units into both conventional and nano-enhanced treatment methods will support Circular Economy (CE) goals, enhancing process economics and sustainability in textile manufacturing by reducing waste and promoting resource recovery.

8.4 *Development of Sustainable and Green Synthesis Routes for Nanomaterials*

Green synthesis of NPs using plant extracts, agro-waste, microbes, and benign solvents is gaining attention as a lower-toxicity, energy-efficient alternative to traditional synthesis methods. This approach is especially important for the textile industry, where dye wastewater treatment can benefit from green NPs. Studies on metal/metal oxide NPs (e.g., Ag, ZnO, Se) prepared by greensynthesis have shown effective dye degradation with reduced environmental footprints [200]. However, further research is needed on scaling these methods, ensuring reproducibility, and standardizing characterization for regulatory acceptance in the textile industry.

8.5 *Policy Frameworks, Safety Regulations, and Standardization Needs*

As nano-enabled dye treatments gain traction in textile wastewater management, robust regulatory policies and safety standards for nanomaterial production, use, and disposal become increasingly critical, especially in technical textile applications. Existing frameworks emphasize life-cycle risk assessments, exposure control, and labeling, but important gaps remain in dose metrics, environmental fate, nanowaste management, and harmonized testing methods [201]. Future regulatory frameworks must address these gaps, ensuring the safe deployment of nano-enabled technologies in wastewater treatment plants treating textile effluents.

8.6 Connecting Laboratory Research with Practical Application

While many promising dye removal technologies for technical textiles have been successfully demonstrated in the laboratory, scaling these innovations to industrial-scale applications requires pilot and demonstration studies. These pilot studies should focus on treating real textile and dyeing effluents with complex compositions. Emerging technologies such as anaerobic reactors, catalytic ozonation, nanofiltration, and membrane distillation show potential for high color/COD removal and water reuse, but challenges like fouling, high costs, and operational robustness must be addressed [202]. To create scale-up strategies, close cooperation between academia, business, and regulators is essential, conduct techno-economic analyses; and implement long-term monitoring for nano-enabled textile wastewater treatment.

9 Conclusion

To create scale-up strategies, close cooperation between academia, business, and regulators is essential, as well as persistent of color and toxic by-products. Methods such as conventional treatment often fail to achieve complete removal or mineralization of dyes, highlighting the need for more advanced solutions. Recent progress in nanomaterial-based technologies has demonstrated high dye removal efficiencies (>90–99%) and improved treatment performance compared with conventional systems. Integrating nanomaterials into hybrid treatment trains offers a promising balance between efficiency, sustainability, and cost when properly designed.

Even though these advantages exist, challenges related to high production costs, scalability, regulatory uncertainty, and potential environmental risks must be addressed through life-cycle assessment, techno-economic analysis, greener synthesis strategies, and optimized system design. Coupling dye removal with resource recovery and circular-economy approaches further enhances sustainability and economic feasibility. Overall, nano-enabled hybrid systems represent a viable and forward-looking pathway for sustainable dye wastewater treatment, supporting environmental protection, public health, and responsible industrial development.

References

1. Diallo MF, Djelassi S, Kumar V (2021) Marketing and globalization: relevance, trends and future research. SAGE Publications, London, pp 2–7. https://doi.org/10.1177/20515707211027326
2. Yuan S, Dai L, Xie M, Liu J, Peng H (2024) Modification optimization and application of graphitic carbon nitride in photocatalysis: current progress and future prospects. Chem Eng Sci 296:120245. https://doi.org/10.1016/j.ces.2024.120245
3. Onu DC, Saidu I, Salawu M, Khan A, Shah NS, Aponte C, Varjani S (2023) Isotherm, kinetics, thermodynamics, recyclability and mechanism of ultrasonic assisted adsorption of methylene blue and lead(II) ions using green synthesized nickel oxide nanoparticles. Environ Nanotechnol Monit Manag 20:100818. https://doi.org/10.1016/j.enmm.2023.100818
4. Jorge AM, Athira KK, Alves MB, Gardas RL, Pereira JFB (2023) Textile dyes effluents: a current scenario and the use of aqueous biphasic systems for the recovery of dyes. J Water Process Eng 55:104125. https://doi.org/10.1016/j.jwpe.2023.104125
5. de Oliveira CRS, Toledo SR, Salahuddin N, Ayad M, Mapossa AB, Carr CM (2025) Dyes and dyeing of textiles: an overview of sustainability issues. In: Sustainable coloration techniques in textiles, pp 1–54. https://doi.org/10.1007/978-981-96-4975-4_1
6. Azanaw A, Birlie A, Teshome B, Jemberie M (2022) Textile effluent treatment methods and eco-friendly resolution of textile wastewater. Case Stud Chem Environ Eng 6:100230. https://doi.org/10.1016/j.cscee.2022.100230
7. Sharma J, Sharma S, Soni V (2021) Classification and impact of synthetic textile dyes on aquatic flora: a review. Reg Stud Mar Sci 45:101802. https://doi.org/10.1016/j.rsma.2021.101802
8. Repon MR, Dev B, Rahman MA, Jurkonienė S, Haji A, Alim MA, Kumpikaitė E (2024) Textile dyeing using natural mordants and dyes: a review. Environ Chem Lett 22:1473–1520. https://doi.org/10.1007/s10311-024-01716-4
9. Sen T, Barrow CJ, Deshmukh SK (2019) Microbial pigments in the food industry challenges and the way forward. Front Nutr 6:7. https://doi.org/10.3389/fnut.2019.00007
10. Dash SK, Kanar S, Pattanaik L, Kamal R, Pradhan A, Dehury RK (2025) Biochar-reinforced metal oxide photocatalysts for the removal of dyes and pharmaceutical pollutants: a review. Ind Biotechnol 12:3767. https://doi.org/10.1177/155090872513767
11. Mahltig B, Böttcher H (2020) Metal complexes as mordants in natural dyeing and printing options and challenges. Molecules 25:3736. https://doi.org/10.3390/molecules25163736
12. Mongkholrattanasit R, Kryštůfek J, Wiener J, Studničková J (2011) Natural dye from eucalyptus leaves and application for wool fabric dyeing by using padding techniques. Nat Dyes 5:57–78. https://doi.org/10.5772/20738
13. Rapiya M, Slayi M, Truter W (2025) A review of phenolic compounds in water bodies of grazing areas. Disc Water 5(1):69. https://doi.org/10.1007/s43832-025-00280-4
14. Holkar CR, Jadhav AJ, Pinjari DV, Mahamuni NM, Pandit AB (2016) A critical review on textile wastewater treatments: possible approaches. J Environ Manage 182:351–366. https://doi.org/10.1016/j.jenvman.2016.07.090
15. Hossain L, Sarker SK, Khan MS (2018) Evaluation of present and future wastewater impacts of textile dyeing industries in Bangladesh. Environ Develop 26:23–33. https://doi.org/10.1016/j.envdev.2018.03.005
16. Thombre N, Patil PS, Yadav A (2025) A short review on water management and reuse in textile industry: a sustainable approach. Disc Water 5:1–17. https://doi.org/10.1007/s43832-025-00215-z
17. Pundir A, Thakur MS, Radha GB, Prakash S, Kumari N, Sharma N, Parameswari E, Senapathy M, Kumar S, Dhumal S, Deshmukh SV, Lorenzo JM, Kumar M (2024) Innovations in textile wastewater management: a review of zero liquid discharge technology. Environ Sci Pollut Res 31:12597–12616. https://doi.org/10.1007/s11356-024-31827-y

18. Connell MG, Rajendran N, Elimelech M, Gilron J, Dunn JB (2024) Analysis of energy, water, land and cost implications of zero and minimal liquid discharge desalination technologies. Nat Water 2(11):1116–1127. https://doi.org/10.1038/s44221-024-00327-1
19. Abdelfattah I, El-Shamy A (2024) Review on the escalating imperative of zero liquid discharge (ZLD) technology for sustainable water management and environmental resilience. J Environ Manage 351:119614. https://doi.org/10.1016/j.jenvman.2023.119614
20. Cáceres-González RA, Hatzell MC (2024) Electrified solar zero liquid discharge: exploring the potential of PV-ZLD in the US. Environ Sci Technol 58:15562–15574. https://doi.org/10.1021/acs.est.4c00494
21. Yaseen D, Scholz M (2019) Textile dye wastewater characteristics and constituents of synthetic effluents: a critical review. Int J Environ Sci Technol 16:1193–1226. https://doi.org/10.1007/s13762-018-2130-z
22. Wang X, Jiang J, Gao W (2022) Reviewing textile wastewater produced by industries: characteristics, environmental impacts, and treatment strategies. Water Sci Technol 85:2076–2096. https://doi.org/10.2166/wst.2022.088
23. Alazaiza MYD, Alzghoul TM, Ramu MB, Amr SSA, Abushammala MFM, Nassani DE (2025) Global perspectives on industrial wastewater management: a bibliometric analysis of research output. J Hazard Mater Adv 17:100567. https://doi.org/10.1016/j.hazadv.2024.100567
24. Ammar EE, Anake WU, Khafaga DS, Stahl H, Ali GA (2025) Advanced nanomaterials for groundwater treatment sustainability. Develop Environ Sci 19:447–469. https://doi.org/10.1016/B978-0-443-33787-1.00015-0
25. Obaideen K, Shehata N, Sayed ET, Abdelkareem MA, Mahmoud MS, Olabi AG (2022) The role of wastewater treatment in achieving sustainable development goals (SDGs) and sustainability guideline. Energy Nexus 7:100112. https://doi.org/10.1016/j.nexus.2022.100112
26. Sharma M, Singh R, Kaur P, Verma A, Patel SK, Kim SH (2025) Microbial fuel cells for azo dye degradation: a perspective review. J Ind Eng Chem 142:45–67. https://doi.org/10.1016/j.jiec.2024.10.012
27. Sombei DC, Mecha CA, Chollom MN (2025) A review of low-cost point-of-use water treatment solutions addressing water access and quality in resource-limited settings. Water 17:1827. https://doi.org/10.3390/w17121827
28. Qureshi F, Ahmad N, Khan SU, Rehman A, Ullah R (2025) Nanomaterials and hydrogen production: a comprehensive review of clean energy strategies, costs, and environmental implications. Results Chem 17:102564. https://doi.org/10.1016/j.rechem.2025.102564
29. Pandey AK, Mishra S, Kumar V, Singh P (2025) Sustainable water management through integrated technologies and circular resource recovery. Environ Sci Water Res Technol 11:1822–1846. https://doi.org/10.1039/D5EW00211A
30. Kurniawan SB, Imron MF, Abdullah SRS, Othman AR (2026) Tackling marine pollution in the blue economy: synergies between wastewater treatment technologies and governmental policies. Mar Pollut Bull 222:118627. https://doi.org/10.1016/j.marpolbul.2025.118627
31. Srinivasan L, Sowjanya S, Chakraborty A, Amrit R, Nurgat MZ, Dutta P, Das P, Das A, Sivarajan S, Wagh S, Mrunal C (2025) Economic, environmental, and social optimization of wastewater management in the context of circular economy. In: Water use efficiency, sustainability and the circular economy. Elsevier, pp 203–230. https://doi.org/10.1016/b978-0-443-26749-9.00011-9
32. Benkhaya S, El-Harfi S, El-Harfi A (2017) Classifications, properties and applications of textile dyes: a review. Appl J Environ Eng Sci 3:311–320
33. Alegbe EO, Uthman U, Olatunde T (2024) A review of history, properties, classification, applications and challenges of natural and synthetic dyes. Heliyon 10:e33646. https://doi.org/10.1016/j.heliyon.2024.e33646
34. Gregory S, Peter A (2000) Dyes and dye intermediates. In: Kirk-othmer encyclopedia of chemical technology, pp 1–66. https://doi.org/10.1002/0471238961.0425051907180507.a01.pub2

35. Christie R, Abel A (2021) Organic pigments: general principles. Phys Sci Rev 6(12):807–834. https://doi.org/10.1515/psr-2020-0187
36. Al-Tohamy R, Ali SS, Li F, Okasha KM, Mahmoud YA-G, Elsamahy T, Jiao H, Fu Y, Sun J (2022) A critical review on the treatment of dye-containing wastewater: ecotoxicological and health concerns of textile dyes and possible remediation approaches for environmental safety. Ecotoxicol Environ Saf 231:113160. https://doi.org/10.1016/j.ecoenv.2021.113160
37. Mukhlish MZ, Khan MR, Islam MS, Nazir MI, Snigdha JS, Akter R, Ahmad H (2020) Decolorization of reactive dyes from aqueous solution using combined coagulation-flocculation and photochemical oxidation (UV/H_2O_2). Sustain Chem Eng 1:51–61
38. López-Grimau V, Gutiérrez-Bouzán C, Valldeperas J (2012) Reuse of the water and salt of reactive dyeing effluent after electrochemical decolorisation. Color Technol 128:36–43. https://doi.org/10.1111/j.1478-4408.2011.00343.x
39. Shabaan OA, Jahin HS, Mohamed GG (2020) Removal of anionic and cationic dyes from wastewater by adsorption using multiwall carbon nanotubes. Arab J Chem 13:4797–4810. https://doi.org/10.1016/j.arabjc.2020.01.010
40. Tolkou AK, Mitropoulos AC, Kyzas GZ (2023) Removal of anthraquinone dye from wastewaters by hybrid modified activated carbons. Environ Sci Pollut Res 30:73688–73701. https://doi.org/10.1007/s11356-023-27550-9
41. Ayed L, Mahdhi A, Cheref A, Bakhrouf A (2011) Decolorization and degradation of azo dye methyl red by an isolated sphingomonas paucimobilis: biotoxicity and metabolites characterization. Desalination 274:272–277. https://doi.org/10.1016/j.desal.2011.02.024
42. Thangaraj S, Bankole PO, Sadasivam SK (2021) Microbial degradation of azo dyes by textile effluent adapted, *Enterobacter hormaechei* under microaerophilic condition. Microbiol Res 250:126805. https://doi.org/10.1016/j.micres.2021.126805
43. Singha K, Pandit P, Maity S, Sharma S (2021) Harmful environmental effects for textile chemical dyeing practice. In: Green chemistry for sustainable textiles. Elsevier, pp 153–164. https://doi.org/10.1016/B978-0-323-85204-3.00005-1
44. Elgarahy AM, Elwakeel KZ, Mohammad SH, Elshoubaky GA (2021) A critical review of biosorption of dyes, heavy metals and metalloids from wastewater as an efficient and green process. Clean Eng Technol 4:100209. https://doi.org/10.1016/j.clet.2021.100209
45. Feng J, Cerniglia CE, Chen H (2012) Toxicological significance of azo dye metabolism by human intestinal microbiota. Front Biosci 4:568. https://doi.org/10.2741/e400
46. Manickam P, Vijay D (2021) Chemical hazards in textiles. In: Muthu SS (ed) Chemical management in textiles and fashion. Elsevier, pp 19–52. https://doi.org/10.1016/B978-0-12-820494-8.00002-2
47. Kishor R, Purchase D, Saratale GD, Ferreira LFR, Bilal M, Bharagava RN (2021) Ecotoxicological and health concerns of persistent coloring pollutants of textile industry wastewater and treatment approaches for environmental safety. J Environ Chem Eng 9:105012. https://doi.org/10.1016/j.jece.2021.105012
48. Oyeniran DO, Sogbanmu TO, Adesalu TA (2021) Antibiotics, algal evaluations and subacute effects of abattoir wastewater on liver function enzymes, genetic and haematologic biomarkers in the freshwater fish, *Clarias gariepinus*. Ecotoxicol Environ Saf 212:111982. https://doi.org/10.1016/j.ecoenv.2021.111982
49. Li F, Zhao K, Ng TSA, Dai Y, Wang CH (2022) Sustainable production of bio-oil and carbonaceous materials from biowaste co-pyrolysis. Chem Eng J 427:131821. https://doi.org/10.1016/j.cej.2021.131821
50. Azimi B, Abdollahzadeh-Sharghi E, Bonakdarpour B (2021) Anaerobic-aerobic processes for the treatment of textile dyeing wastewater containing three commercial reactive azo dyes: effect of number of stages and bioreactor type. Chin J Chem Eng 39:228–239. https://doi.org/10.1016/j.cjche.2020.10.006
51. Wang B, Zhang Y, Li X, Liu H (2022) Biodecolorization and ecotoxicity abatement of disperse dye-production wastewater using *Pycnoporus laccase*. Int J Environ Res Public Health 19:7983. https://doi.org/10.3390/ijerph19137983

52. Velusamy S, Roy A, Sundaram S, Kumar M (2021) A review on heavy metal ions and dye removal through graphene oxide-based adsorption strategies for textile wastewater treatment. Chem Rec 21:1570–1610. https://doi.org/10.1002/tcr.202000146
53. Abrahamse H, Hamblin MR (2016) New photosensitizers for photodynamic therapy. Biochem J 473:347–364. https://doi.org/10.1042/BJ20150942
54. Sharma K, Sharma V, Sharma S (2018) Dye-sensitized solar cells: fundamentals and current status. Nanoscale Res Lett 13:381. https://doi.org/10.1186/s11671-018-2760-6
55. Armitage RA, Lombardi JR, Casadio F (2019) Analysis of the saltzman collection of Peruvian dyes by high-performance liquid chromatography and ambient ionisation mass spectrometry. Heritage Sci 7:1–12. https://doi.org/10.1186/s40494-019-0315-7
56. Liu Y, Xu Z, Wang X, Zhang J (2021) Separation and identification of commercial reactive dyes using hydrophilic interaction liquid chromatography coupled with quadrupole time-of-flight mass spectrometry. Color Technol 137:407–417. https://doi.org/10.1111/cote.12538
57. Śmigiel-Kamińska D, Broda J, Orłowska M (2020) Identification of cotton fibers dyed with reactive dyes for forensic purposes. Molecules 25:5435. https://doi.org/10.3390/molecules 25225435
58. Terán JE, González M, Hernández F (2021) Characterization of synthetic dyes for environmental and forensic assessments using chromatography and mass spectrometry. J Sep Sci 44:387–402. https://doi.org/10.1002/jssc.202000815
59. Whitney AV, Van Duyne RP, Casadio F (2006) Surface-enhanced Raman spectroscopy for identification of historical red lakes and dyestuffs. J Raman Spectrosc 37:993–1002. https://doi.org/10.1002/jrs.1541
60. Zhang YQ, Wei XC, Long JJ (2016) Eco-friendly synthesis and application of disperse-reactive dyes for waterless wool coloration using supercritical carbon dioxide. J Clean Prod 133:746–756. https://doi.org/10.1016/j.jclepro.2016.05.129
61. Krauss M, Singer H, Hollender J (2010) LC–high resolution MS in environmental analysis: from target screening to the identification of unknowns. Anal Bioanal Chem 397:943–951. https://doi.org/10.1007/s00216-010-3608-9
62. Sica VP, Raja S, Adekola O (2020) The role of mass spectrometry and related techniques in the analysis of extractable and leachable chemicals. Mass Spectrom Rev 39:212–226. https://doi.org/10.1002/mas.21591
63. Hassan MM, Carr CM (2018) A critical review on recent advancements of the removal of reactive dyes from dyehouse effluent by ion-exchange adsorbents. Chemosphere 209:201–219. https://doi.org/10.1016/j.chemosphere.2018.06.043
64. Imran M, Crowley DE, Khalid A, Hussain S, Mumtaz MW, Arshad M (2015) Microbial biotechnology for decolorization of textile wastewaters. Rev Environ Sci Biotechnol 14:73–92. https://doi.org/10.1007/s11157-014-9344-4
65. Radhakrishnan S, Rajangam R, Peruran P (2024) Social and environmental impact of natural dyeing. In: Natural dyes and sustainability. Springer, pp 127–152. https://doi.org/10.1007/978-3-031-47471-2_7
66. Chakrabarti A, Chakrabarti D (2017) Research into design for communities (proceedings of ICoRD 2017), vol 1. Springer, Singapore. https://doi.org/10.1007/978-981-10-3518-0
67. Yadav G, Ahmaruzzaman M (2022) Sustainable development of nanomaterials for removal of dyes from water and wastewater. In: Advanced oxidation processes in dye-containing wastewater, vol 1. Springer, pp 167–188. https://doi.org/10.1007/978-981-19-0987-0-8
68. Newman MC (2018) The nature and use of ecotoxicological evidence: natural science, statistics, psychology, and sociology. Academic Press, London
69. Vikrant K, Giri BS, Raza N, Roy K, Kim K-H, Rai BN, Singh RS (2018) Recent advancements in bioremediation of dye: current status and challenges. Biores Technol 253:355–367. https://doi.org/10.1016/j.biortech.2018.01.029
70. Ito T, Adachi Y, Yamanashi Y, Shimada Y (2016) Long-term natural remediation process in textile dye-polluted river sediment driven by bacterial community changes. Water Res 100:458–465. https://doi.org/10.1016/j.watres.2016.05.050

71. Vargas A, Paulino A, Nozaki J (2009) Effects of daily nickel intake on the bio-accumulation, body weight and length in tilapia (*Oreochromis niloticus*). Toxicol Toxicol Environ Chem 91:751–759. https://doi.org/10.1080/02772240903013045
72. Sandhya S (2010) Biodegradation of azo dyes under anaerobic condition: role of azoreductase. In: Biodegradation of azo dyes. UK, London, pp 39–57. https://doi.org/10.1007/698_2009_43
73. Copaciu F (2013) Diffuse water pollution by anthraquinone and azo dyes in environment importantly alters foliage volatiles, carotenoids and physiology in wheat (*Triticum aestivum*). Water Air Soil Pollut 224:1478. https://doi.org/10.1007/s11270-013-1478-4
74. Loomis D, Guha N, Hall AL, Straif K (2018) Identifying occupational carcinogens: an update from the IARC monographs. Occup Environ Med 75:593–603. https://doi.org/10.1136/oemed-2017-104944
75. Samykannu M (2025) Eco-friendly approaches to azo dye removal: the role of microbial azo-reductases. Appl Biochem Biotechnol 197:6358–6376. https://doi.org/10.1007/s12010-025-05343-9
76. Ma G (2025) Endocrine-disrupting effects of 24 colorants used in food contact materials: in vitro and in silico studies. Food Chem Toxicol 200:115410. https://doi.org/10.1016/j.fct.2025.115410
77. National Toxicology Program (2021) Substances delisted from the report on carcinogens. In: Proceedings of the 15th report on carcinogens. National Toxicology Program. https://doi.org/10.22427/NTP-OTHER-1003
78. Gupta BG, Mukhopadhyay R (2025) Heavy metal contamination from textile wastewater and its health impacts: a case study from West Bengal with sustainable remediation approaches. Sci Rep 15:29578. https://doi.org/10.1038/s41598-025-13357-w
79. Hegde VN (2025) Evaluating the impact of closed-loop RO systems on ZLD in textile industry: case study from Tirupur, India. In: Desalination for the environment–clean water and energy congress. https://doi.org/10.24406/publica-4947
80. Mudhoo A (2020) An analysis of the versatility and effectiveness of composts for sequestering heavy metal ions, dyes and xenobiotics from soils and aqueous milieus. Ecotoxicol Environ Saf 197:110587. https://doi.org/10.1016/j.ecoenv.2020.110587
81. Samsami S (2020) Recent advances in the treatment of dye-containing wastewater from textile industries: overview and perspectives. Process Saf Environ Prot 143:138–163. https://doi.org/10.1016/j.psep.2020.05.034
82. Abu-Nada A, Abdala A, McKay G (2021) Removal of phenols and dyes from aqueous solutions using graphene and graphene composite adsorption: a review. J Environ Chem Eng 9:105858. https://doi.org/10.1016/j.jece.2021.105858
83. Muthu SS (2021) Sustainable technologies for textile wastewater treatments. Woodhead Publishing, London
84. Hisada M, Tomizawa Y, Kawase Y (2019) Removal kinetics of cationic azo-dye from aqueous solution by poly-γ-glutamic acid biosorbent: contributions of adsorption and complexation/precipitation to Basic Orange 2 removal. J Environ Chem Eng 7:103157. https://doi.org/10.1016/j.jece.2019.103157
85. Kubra KT, Salman MS, Hasan MN (2021) Enhanced toxic dye removal from wastewater using biodegradable polymeric natural adsorbent. Molecular Liquids 328:115468. https://doi.org/10.1016/j.molliq.2021.115468
86. Saifi A (2021) Complexation of an azo dye by cyclodextrins: a potential strategy for water purification. ACS Omega 6:4776–4782. https://doi.org/10.1021/acsomega.0c05997
87. Hussain J (2024) Degradation of methylene blue under visible-light with copper-doped cobalt oxide nanoparticles. J Taiwan Instit Chem Eng 165:105819. https://doi.org/10.1016/j.jtice.2024.105819
88. Brião GV (2018) Highly efficient and reusable mesoporous zeolite synthetized from a biopolymer for cationic dyes adsorption. Colloids Surf A 556:43–50. https://doi.org/10.1016/j.colsurfa.2018.07.070

89. Minisy I, Salahuddin N, Ayad M (2019) Chitosan/polyaniline hybrid for the removal of cationic and anionic dyes from aqueous solutions. Appl Polym Sci 136:47056. https://doi.org/10.1002/app.47056
90. Bober P, Chalupa J, Holub M, Vlcek J (2020) Conducting polymer composite aerogel with magnetic properties for organic dye removal. Synth Met 260:116266. https://doi.org/10.1016/j.synthmet.2019.116266
91. Ayad MM, Abdul Karim A, Al-Hadeethi Y, Al-Khattab T (2018) Polypyrrole-coated cotton textile as adsorbent of methylene blue dye. Chem Papers 72:1605–1618. https://doi.org/10.1007/s11696-018-0409-0
92. Zheng Y, Zhang J, Sun X, Xu Z (2013) Positively charged thin-film composite hollow fiber nanofiltration membrane for the removal of cationic dyes through submerged filtration. Desalination 328:42–50. https://doi.org/10.1016/j.desal.2013.08.009
93. Rambabu K, Srinivas K, Rao P (2019) Effective treatment of dye polluted wastewater using nanoporous $CaCl_2$ modified polyethersulfone membrane. Process Saf Environ Prot 124:266–278. https://doi.org/10.1016/j.psep.2019.02.015
94. Allahkarami E, Allahkarami E (2025) Filtration and membrane processes. In: Chemical engineering essentials 1: comprehensive chemical engineering, pp 345–391. https://doi.org/10.1002/9781394372348.ch12
95. Alventosa-deLara E, Sanchez-Martin J, Torres F (2012) Ultrafiltration technology with a ceramic membrane for reactive dye removal: optimization of membrane performance. J Hazard Mater 209–210:492–500. https://doi.org/10.1016/j.jhazmat.2012.01.086
96. Wang A-J et al (2020) Electrochemistry-stimulated environmental bioremediation: development of applicable modular electrode and system scale-up. Environ Sci Ecotechnol 3:100050. https://doi.org/10.1016/j.ese.2020.100050
97. Wang A-J, Li Y, Chen Z, Zhang Z (2020) Removing organic matters from reverse osmosis concentrate using advanced oxidation-biological activated carbon process combined with Fe^{3+}/humus-reducing bacteria. Ecotoxicol Environ Saf 203:110945. https://doi.org/10.1016/j.ecoenv.2020.110945
98. Lü Z, Zhang H, Yang Y, Zhang Y (2019) Composite nanofiltration membrane with asymmetric selective separation layer for enhanced separation efficiency to anionic dye aqueous solution. J Hazard Mater 368:436–443. https://doi.org/10.1016/j.jhazmat.2019.01.069
99. Shao H, Zhang Z, Li Y, Zhai Z (2019) Polypropylene composite hollow fiber ultrafiltration membranes with an acrylic hydrogel surface by in situ ultrasonic wave-assisted polymerization for dye removal. J Appl Polym Sci 136:47099. https://doi.org/10.1002/app.47099
100. Fradj AB, Ben Hamida R, Jaziri M, Bouzid J (2020) Removal of azoic dyes from aqueous solutions by chitosan enhanced ultrafiltration. Results Chem 2:100017. https://doi.org/10.1016/j.rechem.2019.100017
101. Tan BH, Teng TT, Omar AM (2000) Removal of dyes and industrial dye wastes by magnesium chloride. Water Res 34:597–601. https://doi.org/10.1016/S0043-1354(99)00151-7
102. Verma AK, Dash RR, Bhunia P (2012) A review on chemical coagulation/flocculation technologies for removal of colour from textile wastewaters. J Environ Manage 93:154–168. https://doi.org/10.1016/j.jenvman.2011.09.012
103. Teh CY, Budiman PM, Shak KPY, Wu TY (2016) Recent advancement of coagulation–flocculation and its application in wastewater treatment. Ind Eng Chem Res 55:4363–4389. https://doi.org/10.1021/acs.iecr.5b04703
104. Wei Q, Zhang Y, Zhang K, Mwasiagi JI, Zhao X, Chow CW, Tang R (2022) Removal of direct dyes by coagulation: adaptability and mechanism related to the molecular structure. Korean J Chem Eng 39:1850–1862. https://doi.org/10.1007/s11814-021-1056-1
105. Deshmukh SM, Dhokpande SR, Sankhe A, Khandekar AR (2025) Effluent wastewater technologies for textile industry: a review. Rev Inorg Chem 45:21–40. https://doi.org/10.1515/revic-2024-0046
106. Aragaw TA, Bogale FM (2023) Role of coagulation/flocculation as a pretreatment option to reduce colloidal/bio-colloidal fouling in tertiary filtration of textile wastewater: a review and future outlooks. Front Environ Sci 11:1142227. https://doi.org/10.3389/fenvs.2023.1142227

107. Rahmoun HB, Benhammou A, Khezzar L, Belhachemi M (2025) Coupling coagulation–flocculation–sedimentation with adsorption on biosorbent (corncob) for the removal of textile dyes from aqueous solutions. Environments 12:201. https://doi.org/10.3390/environments12060201
108. Reyes-Pérez JA, Amaya-Chávez A, Roa-Morales G, Balderas-Hernández P, Pavón Silva TB, Torres-Blancas T, Barrera-Díaz CE (2025) Textile wastewater coupled treatment implementing enhanced ozonation with fenton-like processes and phytoremediation. Catalysts 15:43. https://doi.org/10.3390/catal15010043
109. Zhang Y, Shaad K, Vollmer D, Ma C (2021) Treatment of textile wastewater using advanced oxidation processes—a critical review. Water 13(24):3515. https://doi.org/10.3390/w13243515
110. Afonso C, Sousa CY, Farinon DM, Lopes A, Fernandes A (2024) Electrochemical oxidation of pollutants in textile wastewaters using BDD and Ti-based anode materials. Textiles 4(4):521–529. https://doi.org/10.3390/textiles4040030
111. Rekhate CV, Sen S, Pal D, Prajapati AK (2025) COD and color removal from real textile dyeing effluent using intensification of ozonation and Fenton ($O_3/Fe_2^+/H_2O_2$) processes: optimization using RSM. Disc Electrochem 2:3. https://doi.org/10.1007/s44373-025-00017-x
112. Ajmal A, Majeed I, Malik RN, Idriss H, Nadeem MA (2014) Principles and mechanisms of photocatalytic dye degradation on TiO_2 based photocatalysts: a comparative overview. RSC Adv 4:37003–37026. https://doi.org/10.1039/C4RA06658H
113. Khan S, Ahmad M, Ali M, Khan M (2024) Photocatalytic dye degradation from textile wastewater: a review. ACS Omega 9:21751–21767. https://doi.org/10.1021/acsomega.4c00887
114. Lucas MS, Teixeira AR, Jorge N, Peres JA (2025) Industrial wastewater treatment by coagulation–flocculation and advanced oxidation processes: a review. Water 17:1934. https://doi.org/10.3390/w17131934
115. Qasem NAA, Mohammed RH, Lawal DU (2021) Removal of heavy metal ions from wastewater: a comprehensive and critical review. J Clean Water 4:36. https://doi.org/10.1038/s41545-021-00127-0
116. Ali SS, Al-Tohamy R, Sun J, Wu J, Huizi L (2019) Screening and construction of a novel microbial consortium SSA-6 enriched from the gut symbionts of wood-feeding termite, *Coptotermes formosanus* and its biomass-based biorefineries. Fuel 236:1128–1145. https://doi.org/10.1016/j.fuel.2018.08.117
117. Coria-Oriundo LL, Battaglini F, Wirth SA (2021) Efficient decolorization of recalcitrant dyes at neutral/alkaline pH by a new bacterial laccase-mediator system. Ecotoxicol Environ Saf 217:112237. https://doi.org/10.1016/j.ecoenv.2021.112237
118. Liu W, You Y, Sun D, Wang S, Zhu J, Liu C (2018) Decolorization and detoxification of water-insoluble Sudan dye by *Shewanella putrefaciens* CN32 co-cultured with *Bacillus circulans* BWL1061. Ecotoxicol Environ Saf 166:11–17. https://doi.org/10.1016/j.ecoenv.2018.09.055
119. Guo G, Zhang J, Li X, Wang H (2020) Decolorization and detoxification of azo dye by halo-alkaliphilic bacterial consortium: systematic investigations of performance, pathway and metagenome. Ecotoxicol Environ Saf 204:111073. https://doi.org/10.1016/j.ecoenv.2020.111073
120. Chen G, Zhang H, Li W, Liu Y (2021) Detoxification of azo dye Direct Black G by thermophilic *Anoxybacillus* sp. PDR2 and its application potential in bioremediation. Ecotoxicol Environ Saf 214:112084. https://doi.org/10.1016/j.ecoenv.2021.112084
121. Chen Y-G, Li Z, Zhang J, Wang Y (2021) Impacts of heavy metals and medicinal crops on ecological systems, environmental pollution, cultivation, and production processes in China. Ecotoxicol Environ Saf 219:112336. https://doi.org/10.1016/j.ecoenv.2021.112336
122. Kamali M, Khodaparast Z (2015) Review on recent developments on pulp and paper mill wastewater treatment. Ecotoxicol Environ Saf 114:326–342. https://doi.org/10.1016/j.ecoenv.2014.12.005
123. Srivastava A, Dangi LK, Kumar S, Rani R (2022) Microbial decolorization of Reactive Black 5 dye by *Bacillus albus* DD1 isolated from textile water effluent: kinetic, thermodynamics and decolorization mechanism. Heliyon 8:e08834. https://doi.org/10.1016/j.heliyon.2022.e08834

124. Ikram M, Naeem M, Zahoor M, Rahim A, Hanafiah MM, Oyekanmi AA, Shah AB, Mahnashi MH, Al-Ali A, Jalal NA, Bantun F, Sadiq A (2022) Biodegradation of azo dye methyl red by Pseudomonas aeruginosa: optimization of process conditions. Int J Environ Res Public Health 19:9962. https://doi.org/10.3390/ijerph19169962
125. Tripathi M, Shukla S, Singh R, Singh S, Singh PK, Shukla AK, Maurya S, Pathak S, Chaudhary VK (2024) Physicochemical investigations of textile wastewater and process parameter optimization for biodecolorization of Congo red dye by *Pseudomonas aeruginosa* MT-2 strain. J Pure Appl Microbiol 18:2558–2569
126. Das P, Mehra R, Sharma S, Rai V (2024) Screening different solid supports for Pseudomonas aeruginosa biofilm formation and determining its efficiency for decolorization and degradation of Congo red. Arch Microbiol 206:402. https://doi.org/10.1007/s00203-024-04125-x
127. Mustafa G, Zahid MT, Ali S, Abbas SZ, Rafatullah M (2021) Biodegradation and discoloration of disperse blue-284 textile dye by *Klebsiella pneumoniae* GM-04 bacterial isolate. J King Saud Univ Sci 33:101442. https://doi.org/10.1016/j.jksus.2021.101442
128. El-Awady ME, El-Sheshtawy HS, Abdel-Rahman MA (2024) Exploring the decolorization efficiency and biodegradation mechanisms of different functional textile azo dyes by *Streptomyces albidoflavus* 3MGH. BMC Microbiol 24:210. https://doi.org/10.1186/s12866-024-03347-9
129. Bhandari S, Poudel DK, Marahatha R, Dawadi S (2021) Microbial enzymes used in bioremediation. J Chem 2021:8849512. https://doi.org/10.1155/2021/8849512
130. Pham VHT, Kim J, Chang S, Bang D (2023) Investigating bio-inspired degradation of toxic dyes using potential multi-enzyme producing extremophiles. Microorganisms 11:1273. https://doi.org/10.3390/microorganisms11051273
131. Singh SN (ed) (2015) Microbial degradation of synthetic dyes in wastewaters. Springer, Cham. https://doi.org/10.1007/978-3-319-10942-8
132. Chandra R (ed) (2016) Environmental waste management. CRC Press (Taylor & Francis), Boca Raton. https://doi.org/10.1201/b19243
133. Husain Q (2010) Peroxidase mediated decolorization and remediation of wastewater containing industrial dyes: a review. Rev Environ Sci Biotechnol 9:117–140. https://doi.org/10.1007/s11157-009-9184-9
134. Maurya A, Raj A (2020) Recent advances in the application of biofilm in bioremediation of industrial wastewater and organic pollutants. In: Chowdhary P, Raj A, Verma D, Akhter Y (eds) Microorganisms for sustainable environment and health. Elsevier, pp 81–118. https://doi.org/10.1016/B978-0-12-819001-2.00005-X
135. Flemming H-C, Wingender J, Szewzyk U, Steinberg P, Rice SA, Kjelleberg S (2016) Biofilms: an emergent form of bacterial life. Nat Rev Microbiol 14:563–575. https://doi.org/10.1038/nrmicro.2016.94
136. Zhao Y, Liu D, Huang W, Yang Y, Ji M, Nghiem LD, Trinh QT, Tran NH (2019) Insights into biofilm carriers for biological wastewater treatment processes: current state-of-the-art, challenges, and opportunities. Biores Technol 288:121619. https://doi.org/10.1016/j.biortech.2019.121619
137. Belete B, Desye B, Ambelu A, Yenew C (2023) Micropollutant removal efficiency of advanced wastewater treatment plants: a systematic review. Environ Health Insights 17:11786302231195158. https://doi.org/10.1177/11786302231195158
138. Madan S, Madan R, Hussain A (2022) Advancement in biological wastewater treatment using hybrid moving bed biofilm reactor (MBBR): a review. Appl Water Sci 12:141. https://doi.org/10.1007/s13201-022-01662-y
139. Shah AA, Walia S, Kazemian H (2024) Advancements in combined electrocoagulation processes for sustainable wastewater treatment: a comprehensive review of mechanisms, performance, and emerging applications. Water Res 252:121248. https://doi.org/10.1016/j.watres.2024.121248
140. Ulson de Souza SMAG, Bonilla KAS, Ulson de Souza AA (2010) Removal of COD and color from hydrolyzed textile azo dye by combined ozonation and biological treatment. J Hazard Mater 179(1–3):35–42. https://doi.org/10.1016/j.jhazmat.2010.02.053

141. Rani Y, Gulia J, Lath A, Kumar S, Singh NP, Rani A, Santal S (2025) Aerobic and anaerobic treatment of textile wastewater using membrane technology. In: Aerobic and anaerobic microbial treatment of industrial wastewater. CRC Press, pp 1–19. https://doi.org/10.1201/9781003381327-1
142. Erkan HS, Çağlak A, Soysaloglu A, Takatas B, Onkal Engin G (2020) Performance evaluation of conventional membrane bioreactor and moving bed membrane bioreactor for synthetic textile wastewater treatment. J Water Process Eng 38:101631. https://doi.org/10.1016/j.jwpe.2020.101631
143. Zafar AM, Naeem A, Minhas MA, Hasan MJ, Rafique S, Ikhlaq A (2024) Removal of reactive dyes from textile industrial effluent using electrocoagulation in different parametric conditions of aluminum electrodes. Total Environ Adv 9:200087. https://doi.org/10.1016/j.teadva.2023.200087
144. Yin H, Qiu P, Qian Y, Kong Z, Zheng X, Tang Z, Guo H (2019) Textile wastewater treatment for water reuse: a case study. Processes 7:34. https://doi.org/10.3390/pr7010034
145. Jamaludin M, Tsai Y-C, Lin H-T, Huang C-Y, Choi W, Chen J-G, Sean W-Y (2024) Modeling and control strategies for energy management in a wastewater center: a review on aeration. Energies 17:3162. https://doi.org/10.3390/en17133162
146. Sharma R, Malaviya P (2022) Constructed wetlands for textile wastewater remediation: a review on concept, pollutant removal mechanisms, and integrated technologies for efficiency enhancement. Chemosphere 290:133358. https://doi.org/10.1016/j.chemosphere.2021.133358
147. Haddad M, Abid S, Hamdi M, Bouallagui H (2018) Reduction of adsorbed dyes content in the discharged sludge coming from an industrial textile wastewater treatment plant using aerobic activated sludge process. J Environ Manage 223:936–946. https://doi.org/10.1016/j.jenvman.2018.07.009
148. Zubair M, Roopesh MS, Ullah A (2024) Challenges and prospects: graphene oxide-based materials for water remediation including metal ions and organic pollutants. Environ Sci Nano 11:3693–3720. https://doi.org/10.1039/D4EN00143E
149. El Mchaouri M, Ben Azzouz S, Kharraz J (2025) Engineering TiO_2 photocatalysts for enhanced visible-light activity in wastewater treatment applications. Tetrahedron Green Chem 16:100084. https://doi.org/10.1016/j.tgc.2025.100084
150. Kohli S, Rathee G, Jha I, Phor L, Sable H, Chaudhary V (2025) Strategically engineering 2D MXene-based advanced adsorbents for sustainable wastewater remediation of dyes. Nanoscale 17:20760–20802. https://doi.org/10.1039/D5NR02336J
151. Shafti DM, Dahlan I, Din ATM (2024) A review of the effectiveness of metal–organic frameworks in removing dye effluents. Water Pract Technol 19:4699–4733. https://doi.org/10.2166/wpt.2024.279
152. Prabhu SM, Rane NR, Li X, Otari SV, Girawale SD, Palake AR, Kodam KM, Park Y-K, Ha Y-H, Yadav KK, Khan MA, Jeon B-H (2023) Magnetic nanostructured adsorbents for water treatment: structure-property relationships, chemistry of interactions, and lab-to-industry integration. Chem Eng J 468:143474. https://doi.org/10.1016/j.cej.2023.143474
153. Bhapkar AR, Bhame S (2024) A review on ZnO and its modifications for photocatalytic degradation of prominent textile effluents: synthesis, mechanisms, and future directions. J Environ Chem Eng 12:112553. https://doi.org/10.1016/j.jece.2024.112553
154. Wang C, Jiang R, Yang J, Wang P (2022) Enhanced heterogeneous fenton degradation of organic pollutants by CRC/Fe_3O_4 catalyst at neutral pH. Front Chem 10:892424. https://doi.org/10.3389/fchem.2022.892424
155. Solano R, Maestre D, Mueses M, Herrera A (2023) TiO_2–CuO heterojunction nanoparticles synthesized by green chemistry supported on beach sand granules: optical, morphological and structural characterization. Nano-struct Nano-obj 35:101024. https://doi.org/10.1016/j.nanoso.2023.101024
156. Li G, Ma H, Zhang J, Zhang Q, Li J (2024) Research progress in graphene-based adsorbents for wastewater treatment: preparation, adsorption properties and mechanisms for inorganic and organic pollutants. C 10:78. https://doi.org/10.3390/c10030078

157. Loura N, Rathee K, Dhull R, Singh M, Dhull V (2024) Carbon nanotubes for dye removal: a comprehensive study of batch and fixed-bed adsorption, toxicity, and functionalization approaches. J Water Process Eng 67:106193. https://doi.org/10.1016/j.jwpe.2024.106193
158. Saputra AMA, Piliang AFR, Dellyansyah M, Andriayani GR, Ramadhan RR, Gea S (2024) Synthesis, properties, and utilization of carbon quantum dots as photocatalysts on degradation of organic dyes: a mini review. Catal Commun 187:106914. https://doi.org/10.1016/j.catcom.2024.106914
159. Singh S, Sivaram N, Nath B, Khan NA, Singh J, Ramamurthy PC (2024) Metal organic frameworks for wastewater treatment, renewable energy and circular economy contributions. NPJ Clean Water 7:124. https://doi.org/10.1038/s41545-024-00408-4
160. Ge S, Wei K, Peng W, Huang R, Akinlabi E, Xia H, Shahzad MW, Zhang X, Xu BB, Jiang J (2024) A comprehensive review of covalent organic frameworks (COFs) and their derivatives in environmental pollution control. Chem Soc Rev 53:11259–11302. https://doi.org/10.1039/D4CS00521J
161. Khan S, Ajmal S, Hussain T, Ur Rahman M (2025) Clay-based materials for enhanced water treatment: adsorption mechanisms, challenges, and future directions. J Umm Al-Qura Univ Appl Sci 11:219–234. https://doi.org/10.1007/s43994-023-00083-0
162. Richard J, Vashishtha A, Phimphachanh A, Rydzek G, Lacroix-Desmazes P, Marcotte N, Gérardin C (2024) Mesoporous silica for sustainable dye removal: fast and reversible adsorption from ordered mesopores densely functionalized with polymers. Microp Mesop Mater 379:113254. https://doi.org/10.1016/j.micromeso.2024.113254
163. Moosavi S, Lai CW, Gan S, Zamiri G, Akbarzadeh Pivehzhani O, Johan MR (2020) Application of efficient magnetic particles and activated carbon for dye removal from wastewater. ACS Omega 5:20684–20697. https://doi.org/10.1021/acsomega.0c01905
164. Qi L, Wang S, Liu Y, Zhao P, Tian J, Zhu B, Zhang S, Xie W, Yu H (2024) Facile preparation of magnetically separable Fe_3O_4/ZnO nanocomposite with enhanced photocatalytic activity for degradation of Rhodamine B. Nanomaterials 14:926. https://doi.org/10.3390/nano14110926
165. González-Fernández C, Bringas E, Rivero MJ, Ortiz I (2024) Revealing the role of magnetic materials in light-driven advanced oxidation processes: enhanced degradation of contaminants and facilitated magnetic recovery. Front Chem Eng 6:1430773. https://doi.org/10.3389/fceng.2024.1430773
166. Kirubakaran D, Abdul Wahid JB, Karmegam N, Jeevika R, Sellapillai L, Rajkumar M, SenthilKumar KJ (2025) A comprehensive review on the green synthesis of nanoparticles: advancements in biomedical and environmental applications. Biomed Biomater Dev 4:388–413. https://doi.org/10.1007/s44174-025-00295-4
167. Ikram Z, Azmat E, Perviaz M (2024) Degradation efficiency of organic dyes on CQDs as photocatalysts: a review. ACS Omega 9:10017–10029. https://doi.org/10.1021/acsomega.3c09547
168. Gengan S, Ananda Murthy HC, Sillanpää M, Nhat T (2022) Carbon dots and their application as photocatalyst in dye degradation studies-mini review. Results Chem 4:100674. https://doi.org/10.1016/j.rechem.2022.100674
169. Yadav M, Yadav A, Fernandes R, Popat Y, Orlandi M, Dashora A, Kothari DC, Miotello A, Ahuja BL, Patel N (2017) Tungsten-doped TiO_2/reduced graphene oxide nano-composite photocatalyst for degradation of phenol: a system to reduce surface and bulk electron-hole recombination. J Environ Manage 203:364–374. https://doi.org/10.1016/j.jenvman.2017.08.010
170. Muradov MB, Mammadyarova SJ, Eyvazova GM, Balayeva OO, Aliyeva G, Hasanova I, Melikova SZ, Musayeva N, Sadigov N, Abdullayev MI (2024) Synthesis of $CuxCo_3-xO_4$ nanoparticles by a sonochemical method and characterization of structural and optical properties and photocatalytic activity for the degradation of methylene blue. RSC Adv 14:1082–1093. https://doi.org/10.1039/D2RA08060E
171. Bissenova M, Idrissov N, Kuspanov Z, Daulbayev C (2025) Hybrid adsorption–photocatalysis composites: a sustainable route for efficient water purification. Mater Renew Sustain Energy 14:1–26. https://doi.org/10.1007/s40243-025-00319-5

172. Mondal A, Prabhakaran A, Gupta S, Subramanian VR (2021) Boosting photocatalytic activity using reduced graphene oxide (RGO)/semiconductor nanocomposites: issues and future scope. ACS Omega 6:8734–8743. https://doi.org/10.1021/acsomega.0c06045
173. Kumari H, Sonia S, Ranga R, Chahal S, Devi S, Sharma S, Kumar S, Kumar P, Kumar S, Kumar A, Parmar R (2023) A review on photocatalysis used for wastewater treatment: dye degradation. Water Air Soil Pollut 234:349. https://doi.org/10.1007/s11270-023-06359-9
174. Sales DA, Lima PNS, Silva LS, Marques TMF, Gusmão SBS, Ferreira OP, Ghosh A, Guerra Y, Morais AÍS, Bezerra RDS, Silva-Filho EC, Viana BC (2022) Amino-functionalized titanate nanotubes: pH and kinetic study of a promising adsorbent for acid dye in aqueous solution. Materials 15:6393. https://doi.org/10.3390/ma15186393
175. Ali H, Ismail AM (2023) Fabrication of magnetic Fe_3O_4/polypyrrole/carbon black nanocomposite for effective uptake of Congo red and methylene blue dye: adsorption investigation and mechanism. J Polym Environ 31:976–998. https://doi.org/10.1007/s10924-022-02663-3
176. Vassileva P, Voykova D, Kichukova D, Lazarova T, Atanasova G, Kovacheva D, Spassova I (2025) A highly efficient graphene-based material for the removal of cationic dyes from aqueous solutions. Materials 18:853. https://doi.org/10.3390/ma18040853
177. Tan SN, Yuen ML, Ramli RA (2025) Photocatalysis of dyes: operational parameters, mechanisms, and degradation pathway. Anal Chem 12:100230. https://doi.org/10.1016/j.greeac.2025.100230
178. Pasindu V, Yapa P, Dabare S, Munaweera I (2025) Multifunctional transition metal oxide/graphene oxide nanocomposites for catalytic dye degradation, renewable energy, and energy storage applications. RSC Adv 15:33162–33186. https://doi.org/10.1039/D5RA04806K
179. Padmanabhan NT, Thomas N, Louis J, Mathew DT, Ganguly P, John H, Pillai SC (2021) Graphene coupled TiO_2 photocatalysts for environmental applications: a review. Chemosphere 271:129506. https://doi.org/10.1016/j.chemosphere.2020.129506
180. Tong S, Zhou J, Ding L, Zhou C, Liu Y, Li S, Meng J, Zhu S, Chatterjee S, Liang F (2022) Preparation of carbon quantum dots/TiO_2 composite and application for enhanced photodegradation of rhodamine B. Colloids Surf A Physicochem Eng Aspects 648:129342. https://doi.org/10.1016/j.colsurfa.2022.129342
181. Bakbolat B, Daulbayev C, Sultanov F, Beissenov R, Umirzakov A, Mereke A, Bekbaev A, Chuprakov I (2020) Recent developments of TiO_2-based photocatalysis in the hydrogen evolution and photodegradation: a review. Nanomaterials 10:1790. https://doi.org/10.3390/nano10091790
182. Xuan S, Jiang W-Q, Gong X-L, Hu Y, Chen Z-Y (2009) Magnetically separable Fe_3O_4/TiO_2 hollow spheres: fabrication and photocatalytic activity. J Phys Chem C 113:553–558. https://doi.org/10.1021/jp8073859
183. Aziz SQ, Ismael SO, Salih KKM (2025) Nanotechnology in wastewater treatment: efficiency review, cost analysis, and reusing. Polytechn J 15:5
184. Mpongwana N, Rathilal S (2022) A review of the techno-economic feasibility of nanoparticle application for wastewater treatment. Water 14:1550. https://doi.org/10.3390/w14101550
185. Al-Rajhi AMH, Abdelghany TM (2025) Innovative strategies for wastewater treatment: Harnessing green technologies for sustainable resource recovery: a review. Front Environ Sci 13:1696485. https://doi.org/10.3389/fenvs.2025.1696485
186. Gul S, Nazar M, Sharif MN, Tariq R, Fatima I, Sarfraz A, Zaheer MA, Sarfraz R, Mustafa Z (2025) Advanced nanotechnology in wastewater treatment: investigating the role of nanoparticles in pollutant removal, water recovery, and environmental sustainability. Schol J Eng Technol 13(5):331–356
187. Dalvand A, Gholami M, Joneidi A, Mahmoodi NM (2011) Dye removal, energy consumption and operating cost of electrocoagulation of textile wastewater as a clean process. Clean: Soil, Air, Water 39:665–672
188. Yang X, López-Grimau V, Vilaseca M, Crespi M (2020) Treatment of textile wastewater by CAS, MBR, and MBBR: a comparative study from technical, economic, and environmental perspectives. Water 12:1306. https://doi.org/10.3390/w12051306

189. Saad MS, Joe NC, Shuib HA, Wirzal MDH, Putra ZA, Khan MR, Busquets R (2022) Techno-economic analysis of an integrated electrocoagulation-membrane system in treatment of palm oil mill effluent. J King Saud Univ Sci 34:102015. https://doi.org/10.1016/j.jksus.2022.102015
190. Ribeiro JP, Sarinho L, Nunes MI (2024) Application of life cycle assessment to Fenton processes in wastewater treatment: a review. J Water Process Eng 57:104692. https://doi.org/10.1016/j.jwpe.2023.104692
191. Mugaishudeen G, Ponnusamy V, Subramanian A (2025) Life Cycle Assessment for the treatment of textile dye wastewater: integrated jet loop reactor and ozonation versus conventional process. Res Sq 3:35
192. Dogan K, Turkmen BA, Arslan-Alaton I, Germirli Babuna F (2024) Life cycle assessment as a decision-making tool for photochemical treatment of iprodione fungicide from wastewater. Water 16:1183. https://doi.org/10.3390/w16081183
193. Silva Santos DH, Xiao Y, Chaukura N, Hill JM, Selvasembian R, Zanta CLPS, Meili L (2022) Regeneration of dye-saturated activated carbon through advanced oxidative processes: a review. Heliyon 8:e10205. https://doi.org/10.1016/j.heliyon.2022.e10205
194. Santos S, Botelho C, Pintor A (2022) Adsorbents for water and wastewater treatment and resource recovery. MDPI Basel. https://doi.org/10.3390/books978-3-0365-4721-3
195. Qian W, Wei W, Liu Z, Zhang L, Li Z (2024) Dyeing sludge-derived biochar for efficient removal of malachite green from dyeing wastewater. Water Emerg Contam Nanoplast 3:25
196. Vergili I, Demirbas A, Karagoz S, Al-Ghamdi A (2012) Techno-economic analysis of textile dye bath wastewater treatment by integrated membrane processes under the zero liquid discharge approach. Resour Conserv Recycl 58:25–35. https://doi.org/10.1016/j.resconrec.2011.10.005
197. Alshehri HS (2025) Recent advances in nanomaterial-based wastewater treatment: a sustainable approach. Polish J Environ Stud 4:5261
198. Baarimah AO, Bazel MA, Alaloul WS, Alazaiza MYD, Al-Zghoul TM, Almuhaya B, Khan A, Mushtaha AW (2024) Artificial intelligence in wastewater treatment: research trends and future perspectives through bibliometric analysis. Case Stud Chem Environ Eng 10:100926. https://doi.org/10.1016/j.cscee.2024.100926
199. Jahan N, Tahmid M, Shoronika AZ, Fariha A, Roy H, Pervez MN, Cai Y, Naddeo V, Islam MS (2022) A comprehensive review on the sustainable treatment of textile wastewater: zero liquid discharge and resource recovery perspectives. Sustainability 14:15398. https://doi.org/10.3390/su142215398
200. Roopa MC, Thirumala S, Kallimani S, Manohara BM, Basavarajaiah SM (2025) Green synthesis of nanoparticles for enhanced wastewater treatment and other applications: a review and future perspectives. Next Mater 8:100664. https://doi.org/10.1016/j.nxmate.2025.100664
201. Isibor PO, Devi G, Enuneku AA (2024) Regulations and policy considerations for nanoparticle safety. In: Isibor PO, Devi G, Enuneku AA (eds) Environmental nanotoxicology: combatting the minute contaminants. Springer, Cham, pp 295–316. https://doi.org/10.1007/978-3-031-54154-4-14
202. Kuchtová G, Herink P, Herink T, Chýlková J, Mikulášek P, Dušek L (2023) From lab-scale to pilot-scale treatment of real wastewater from the production of rayon fiber. Process Saf Environ Prot 171:834846. https://doi.org/10.1016/j.psep.2023.01.059

Innovative Fibers for Sustainability in Fashion: Luxury Versus Fast Fashion Consumer Responses

Maria Antonietta Raimondo and Carmela Donato

Abstract The fashion industry, one of the most resource-intensive sectors globally, is increasingly adopting sustainable practices across its value chain, from material selection to production and distribution. Among emerging innovations for sustainability, textiles derived from food waste, such as Orange Fiber made from orange production, offer both environmental and aesthetic benefits, distinguishing themselves from traditional recycled fibers like those from plastic bottles. This chapter investigates how the origin of textile (Orange Fiber versus recycled plastic fiber) affects consumers' willingness to buy, considering the moderating role of brand type (luxury versus fast fashion). Through an experimental study, results indicate that Orange Fiber generates higher willingness to buy than recycled plastic fiber for luxury brands, while the advantage diminishes for fast fashion brands. The effect of textile origin on willingness to buy is mediated by perceived coolness, reflecting the symbolic value consumers associate with innovative sustainable materials. These findings highlight that combining novel technical textiles for sustainability with strong brand positioning can enhance consumer adoption, aligning ethical responsibility with perceived product desirability.

Keywords Sustainability · Technical textiles · Orange fiber · Coolness · Luxury · Fast fashion · Brand positioning · Willingness to buy

---

M. A. Raimondo (✉)
Department of Business Administration and Law, University of Calabria, Rende, Italy
e-mail: ma.raimondo@unical.it

C. Donato
Department of Economics, University of Roma Tre, Rome, Italy
e-mail: carmela.donato@uniroma3.it

S. S. Muthu (ed.), *Technical Textiles for Sustainability*, Sustainable Textiles: Production, Processing, Manufacturing & Chemistry, https://doi.org/10.1007/978-3-032-29974-1_3

1 Introduction

The fashion industry, recognized as one of the most resource-intensive sectors globally, has increasingly embraced sustainable practices across its value chain, from design and material selection to production and distribution.[1] Sustainable fashion encompasses multiple strategies, including the use of recycled materials, the adoption of low-impact fibers, and circular business models involving resale, rental, and repair. In this context, consumers increasingly favor brands that communicate sustainability through their products and practices, as ethical considerations influence perceived value and brand loyalty [1, 2].

An emerging area of innovation for sustainability in the fashion industry involves technical textiles derived from biological sources, including mushrooms, pineapples, and orange waste. Compared to more traditional sustainable fibers, such as recycled fibers produced from waste materials like plastic bottles or old fabrics, these more innovative fibers can reduce environmental impact, while also enhancing product desirability through the innovative processes behind their production. Among these, Orange Fiber stands out as the first textile made from orange waste, combining ethical and aesthetic value while supporting sustainable resource use.

Despite the growing attention to sustainable materials in the fashion industry, the comparative evaluation of food-waste-derived fibers versus recycled plastics remains underexplored, particularly regarding their consumer appeal across different brand types.

This chapter presents an experimental study examining how textile origin (Orange Fiber versus recycled plastic fiber) influences consumer's willingness to buy, with attention to how these reactions vary according to brand positioning (luxury versus fast fashion). The study demonstrates that Orange Fiber generates higher willingness to buy than recycled plastic fiber when associated with luxury brands. The advantage of Orange Fiber over recycled plastic fiber is less pronounced for fast fashion brands. The relationship between textile origin and willingness to buy, moderated by the brand type, is mediated by perceived coolness [3], which reflects the greater symbolic value consumers associate with more innovative sustainable materials. These findings suggest that the combination of innovative sustainable fibers and strong brand positioning can enhance consumer adoption, bridging environmental responsibility and perceived product desirability.

This study contributes to the literature on technical textiles for sustainability in the fashion industry, particularly through the comparison of luxury and fast fashion brands, and expands current knowledge on the concept of brand coolness. It also provides practical guidance for managers in the fashion industry.

The chapter begins with the theoretical background, including a review of the relevant literature and the development of the conceptual model. It then presents the methodology and results of the experimental study. The discussion section outlines the implications of the findings for theory, managerial practice, and future research.

[1] https://www.mckinsey.com/featured-insights/mckinsey-live/webinars/sustainable-fashion-how-the-fashion-industry-can-urgently-act-to-reduce-its-greenhouse-gas-emission.

2 Theoretical Background: Technical Textiles for Sustainability and Consumer Responses

In this section, we discuss technical textiles for sustainability by reviewing the literature on sustainable consumption, sustainable fashion, sustainability and innovation, sustainable production in the textile industry, and sustainable textile fibers. We then present the conceptual model and develop the research hypotheses concerning consumer responses to different types of sustainable technical textiles in fashion.

2.1 Literature Review

Sustainable Consumption

Over the past decade, the rising consumption of goods and services has produced a profound negative impact on the environment, leading to substantial degradation of natural ecosystems. Among the most pressing consequences are accelerated global warming, increased pollution levels, and a significant decline in numerous plant and animal species [4].

This alarming situation has prompted increased awareness in many countries, which have begun adopting measures to reduce the environmental footprint of production and commercial activities. From this awareness emerged the concept of "*sustainable development*", commonly defined as development that meets present needs without compromising the ability of future generations to meet their own [5]. By emphasizing the promotion of sustainability and supporting development models that minimize environmental harm, sustainable development fosters *green consumption* (or sustainable consumption) and *eco-innovation*, integrating environmentally responsible practices across all stages of production and service delivery [6].

Sustainable consumption, in turn, refers to the environmentally responsible behavior of consumers, who consider the ecological impact of products not only at the point of purchase but also throughout their use and disposal, as well as in their consumption of eco-friendly services [7]. The concept of green consumption has gained increasing prominence, with a growing number of consumers seeking to reduce their environmental footprint through responsible purchasing and disposal choices.

One of the earliest definitions of sustainable consumption emerged during the 1994 Oslo Symposium, describing it as "the use of goods and services that respond to basic needs and bring a better quality of life, while minimizing the use of natural resources and toxic materials, as well as the emissions of waste and pollutants over the life cycle of the service or product, so as not to jeopardize the needs of future generations" [8].

Building on this definition, sustainable consumption refers to the purchase and use of products and services that reduce environmental, social, and economic impacts, thereby safeguarding the planet's natural resources. Moreover, it contributes to mitigating climate change and improving the quality of life for individuals and communities [9].

The growing importance of sustainable consumption is primarily driven by global population growth and globalization, which place increasing pressure on natural resources and demand for goods and services [10]. At the same time, the increased awareness of environmental and social issues is motivating consumers to seek products and services that are both environmentally friendly and ethically produced [11].

Sustainable consumption has increasingly spread across industries, with the *fashion sector* showing a marked commitment to reducing its negative environmental impact [12]. Many brands are adopting greener practices and materials, optimizing production chains and reducing waste to align with sustainability goals.

Consumers and Sustainability: Challenges and Opportunities for Companies

The COVID-19 pandemic had a significant impact on society and the global economy, affecting various sectors in different ways. This unprecedented situation led to a series of consequences for individuals' daily lives and community dynamics, resulting in a radical change in both corporate strategy and consumer behavior [13, 14].

One of the most notable changes observed has been the growing interest in purchasing sustainable products. Consumers have become more cautious and selective in their purchasing decisions, increasingly favouring goods produced in an eco-friendly manner and with a reduced environmental footprint [15].

A study conducted by GfK[2] confirms this trend: at the European level, one in three households has stopped buying certain products because of their negative environmental impact. However, beyond the pandemic, this shift reflects a broader transformation that has been taking place for decades in consumer habits, choices, and tastes. Society has become increasingly attentive to products that are sustainable from an environmental, ethical, and social perspective [16]. This evolution in purchasing behavior offers substantial opportunities for companies capable of offering environmentally responsible products and services.

Firstly, firms that operate sustainably can enhance their corporate image [17] and reputation [18], distinguishing themselves from competitors and earning customer trust [19]. As consumers become more attentive to product origins and production or distribution practices, they are increasingly willing to pay more for sustainable goods and to reward companies that demonstrate compliance with environmental standards [20].

In addition, companies that invest in sustainability can reduce costs and increase efficiency through sustainable practices such as the use of renewable energy, recycling, and waste reduction. These practices can lead to significant operational savings while also strengthening corporate reputation [18].

However, companies must also face challenges when attempting to meet consumer needs in an environmentally sustainable way. Although communicating sustainability

[2] https://www.gfk.com/it/stampa/un-terzo-delle-famiglie-italiane-evita-di-acquistare-prodotti-non-sostenibili.

aspects in the production process, especially in the apparel sector, can be advantageous, many firms engage in greenwashing, that is, misleading advertising about their environmental credentials [21]. Greenwashing occurs when an organization disseminates positive information about its sustainability performance while being aware of shortcomings in that area.

As a result, consumers often distrust sustainability claims, since they cannot easily verify their authenticity [22]. One consequence of greenwashing is that companies emphasizing their social or environmental credentials are initially viewed with skepticism. Building trustful relationships therefore becomes a long-term process that requires time to establish, strengthen, and maintain [23].

Consequently, one of the main challenges for companies is to communicate their sustainability practices effectively, in order to convince even the most skeptical consumers of the credibility of their environmental claims [24]. Firms must thus be able to demonstrate their efforts concretely through certifications and internationally recognized standards. Moreover, companies need to take into account consumers' diverse opinions and preferences regarding sustainability and environmental or social issues to remain competitive in the market.

They should also consider the long-term impact of their business decisions, addressing the entire life cycle of their products, from production to disposal, to minimize the environmental footprint at each stage [25].

In conclusion, while sustainability poses a significant challenge, it also offers remarkable opportunities for corporate growth and innovation.

Sustainable Fashion

At first glance, *fashion and sustainability* might seem like intrinsically contradictory concepts, as the former is defined by hedonism and short product life cycles, especially in fast fashion [26], while the latter emphasizes ethics, durability, and product reuse [27]. However, in recent years, there has been a substantial convergence between the fashion industry and ethical principles. According to McNeill and Snowdon [28], consumers have become increasingly aware of the consequences of textile and apparel production, taking into account broader ethical implications when making purchasing decisions.

As a consequence, the fashion industry has shown growing concern for reducing the environmental impact of its products and production processes [29]. In this context, the concept of "ethical fashion" has emerged. Although difficult to define due to the lack of a universally accepted industry standard, ethical fashion represents an innovative approach to "responsible fashion". It involves a growing number of ethical clothing companies seeking to engage especially young consumers by offering fashionable garments [30].

The idea behind ethical fashion is to produce garments responsibly, while ensuring good labor standards and improved working conditions, promoting sustainable business models in the countries of production, and prioritizing the use of organic materials to minimize environmental impact [31]. For these reasons, ethical fashion can be described as trend-conscious clothing that integrates fair-trade principles, promotes adequate working conditions for employees, and uses biodegradable and organic materials to reduce environmental harm [32].

Another concept used to describe the environmental impact of the fashion industry is "sustainable fashion." These two terms (i.e., ethical fashion and sustainable fashion) are often used interchangeably, but they carry slightly different meanings: ethical fashion focuses primarily on the ethical aspects of clothing production, whereas sustainable fashion emphasizes environmental aspects. Nevertheless, the two are closely related: a healthy and long-lasting fashion industry must be both sustainable and ethical, as these two dimensions cannot exist independently.[3]

Sustainable fashion developed as part of the slow fashion movement [33], which first appeared in the 1960s, when many consumers began to notice the harmful effects of industrial clothing production on the environment and demanded a transformation of industrial practices [34]. Initially, sustainable fashion was not perceived positively, but this began to change with the anti-fur campaigns of the 1980s and 1990s, followed by a growing interest in ethical clothing at the end of the 1990s [35].

Today, the slow fashion and sustainable fashion movements aim to empower workers across the entire supply chain, using practices such as upcycling, recycling, and traditional production techniques, while incorporating renewable raw materials. This requires a radical transformation of current fashion industry practices centered on growth. Consequently, systemic approaches, infrastructures, and production efficiency must be rethought [36].

The central principle of slow fashion and sustainable fashion lies in a balanced approach to garment production, promoting long-term relationships between producers and consumers, supporting local manufacturing, and fostering transparency [12].

It is therefore essential that companies demonstrate their commitment to sustainability through the adoption of various sustainable practices: the use of recycled or organic materials, waste reduction, the implementation of ethical and transparent supply chains, and the reduction of the environmental impact of production and transportation [37].

Promoting sustainability in the fashion sector is crucial for ensuring the well-being of the planet and future generations. Sustainable and slow fashion, by encouraging ethical and responsible approaches, can be seen as a long-term solution for the fashion industry capable of balancing economic and environmental priorities [34].

Consequently, sustainable and ethical fashion have become key priorities for the fashion industry, as also demonstrated by a 2022 Statista[4] survey, which found that 35% of U.S. consumers are willing to pay significantly more for eco-friendly or sustainable clothing.

From this evidence, it is clear that consumers are playing an active role in reducing the environmental impact of the fashion industry. This trend is pushing companies

[3] https://www.econyl.com/blog/community/the-difference-between-sustainable-and-ethical-fashion-and-why-we-need-both/.

[4] https://www.statista.com/forecasts/1235883/willingness-to-pay-for-sustainable-fashion-in-the-us.

to rethink their strategies and invest in sustainable initiatives to meet public expectations while minimizing their environmental and social footprint [38]. This transformation involves not only established brands but also numerous start-ups committed to creating sustainable fashion and gaining increasing public support.

The ultimate goal is to build a more ethical, fair, and eco-friendly fashion industry capable of meeting consumer needs without harming the environment or the people involved in production. Today, sustainability is no longer an option but a necessity for companies that wish to thrive in the long term and maintain the trust of their customers.

Fashion Innovation and Sustainability

Sustainability in the fashion industry is not limited to the use of eco-friendly materials and responsible production practices; it also involves the development of innovations and advanced technologies that enable the achievement of environmental goals. Consequently, sustainability and innovation are closely interrelated and continuously evolving concepts.

Sustainable innovation, also known as *green innovation* or *eco-innovation*, refers to the reduction of environmental impact or the pursuit of specific ecological development goals through the introduction of new ideas, behaviors, products, and processes. These include the creation of new green products, the adoption of environmental protection technologies, and the consideration of sustainable development in the exploration of new markets, as well as the implementation of new production systems and management processes [39, 40].

Innovation contributes significantly to sustainability in multiple ways [41]:

- by fostering research and development of innovative materials and advanced technologies that aim to reduce the environmental impact of goods and services;
- by encouraging the use of renewable energy and eco-friendly transportation systems to mitigate the environmental effects of human activity;
- a by leveraging digitalization, which leads to reduced use of paper and materials, thus decreasing resource consumption and CO_2 emissions.

Moreover, innovation creates new business opportunities and enhances the competitiveness of firms that, by investing in green innovation, can improve their brand image, attract new customers, and increase operational efficiency, thereby contributing to economic growth and sustainable development [42].

According to Pereira and Vence [43], the success of sustainable innovation depends on company size, age, and competencies. Large firms are more likely to engage in sustainable innovation due to greater resources and economies of scale, whereas small and medium-sized enterprises (SMEs) often lack green skills, innovation capacity, and a green image. Therefore, SMEs must invest in the development of green competencies to bridge the gap with larger firms and remain competitive.

Both SMEs and large corporations should implement change management and organizational transformation strategies to overcome inertia and promote sustainable practices. Collaboration between companies, institutions, and research centers can also play a key role in driving green innovation and spreading environmentally friendly technologies.

A prominent example of sustainable innovation is the circular economy, a model that has gained increasing prominence in the current economic and environmental landscape. The circular economy aims to minimize waste, both in terms of energy and resources, by promoting reduction, reuse, and recycling to create a sustainable cycle of production and consumption [44]. Unlike the traditional linear economy based on the "take-make-use-dispose" model, which generates significant waste and pollution [45], the circular economy seeks to create value through waste reduction and resource efficiency. Concrete examples include material recycling, product repair and reuse, life extension through redesign, resource sharing, and collaborative business models [44]. This sustainable economic vision helps reduce operating costs and greenhouse gas emissions [46], while also enhancing brand image, strengthening consumer-brand identification, and increasing loyalty [47].

Sustainable innovation is also becoming increasingly central in the fashion industry, traditionally associated with high environmental impact due to intensive water and chemical use, as well as large quantities of textile waste [48]. The sector has also faced criticism for unethical labor conditions in its supply chains [49]. A clear example is the fast fashion brand Shein, which has been criticized not only for pollution caused by its activities but also for poor labor conditions. A recent Channel 4 investigation revealed the harsh working conditions of its employees, who are paid only 4 cents per garment, producing at least 500 pieces per day, with 18-h shifts.[5] Shein is just one among many fashion brands accused of questionable practices.

Notably, several other brands are now striving to reduce their environmental impact by adopting sustainable innovation based on circular economy principles [50]. One of the most promising innovations involves the use of textile fibers made from organic or food waste, such as fruit, corn, or milk by-products [51].

For instance, H&M, in its *Conscious Exclusive* collection, collaborated with the Italian company Vegea, creating shoes and bags made from grape-based textile fibers produced using waste from the wine industry.[6] Similarly, Salvatore Ferragamo, in 2017, launched a collection in collaboration with Orange Fiber, a Sicilian company that produces sustainable fabrics from orange juice by-products.

Other companies are leveraging digital technologies to improve supply chain transparency. For example, Provenance, a social enterprise, is developing a blockchain-based system to track supply chains and product life cycles.[7]

In conclusion, innovation and sustainability are deeply interconnected, and their combined development across industries is fostering a more sustainable and socially responsible future. Adopting an innovative and sustainable approach can help companies create long-term value, enhance their reputation, and meet the growing expectations of consumers who increasingly demand environmentally responsible solutions [19].

[5] https://forbes.it/2022/10/21/shein-sfruttamento-lavoro-channel-4-fast-fashion/.

[6] https://www.vegeacompany.com/vegea-for-hm-conscious-exclusive/.

[7] https://www.provenance.org/whitepaper.

The Challenge of Sustainable Production in Textile Industry

In recent years, as illustrated in the previous sections, the textile industry has been widely criticized due to its intensive use of natural resources and the negative environmental and social impacts associated with garment production and disposal. For this reason, the sector is directing its efforts toward a circular economy for textiles, aimed at protecting resources, the environment, and people through the development of innovative recycling technologies [52].

In this context, the use of sustainable textile fibers can represent an effective solution to mitigate the environmental damage caused by the fashion industry, as they allow the production of new fabrics and garments with a lower impact compared to traditional fibers [53]. Indeed, traditional fibers such as cotton and polyester require large amounts of water, energy, and chemical products for their production, moreover, most of them end up in landfills at the end of their life cycle, contributing to the growing waste crisis [54]. This is not the case with certain sustainable textile fibers, which require less water, energy, and chemicals to produce and can be recycled or composted at the end of their life cycle. For this reason, their production can enhance a company's reputation and competitive position [55]. However, to expand the production of sustainable textile fibers, companies must adopt increasingly sustainable production techniques, invest in research and development to find new solutions, and collaborate with other firms and industry stakeholders.

In particular, three techniques can be used for the production of sustainable textile fibers: recycling, upcycling, and the use of new technologies aimed at creating fibers from organic and food waste.[8]

Under Directive 2008/98/EC,[9] recycling is defined as "any recovery operation by which waste materials are reprocessed into products, materials, or substances whether for the original or other purposes". In the fashion industry, recycling consists of collecting production scraps and other unused fabrics or materials to transform them into new textile products. The recycling process can be carried out mechanically or chemically and involves sorting and separating fabrics according to fiber composition and quality [56]. Textile fiber recycling represents an important opportunity to reduce environmental impacts in the fashion industry, as the various recycling methods available offer sustainable solutions for resource conservation and the creation of new products from existing materials [57].

Upcycling, refers to the reuse of discarded objects or materials to create a product of higher quality or value than the original [58, 59].

Recycling and upcycling are two sustainability-related concepts that contribute to reducing the environmental impact of waste. Although they share similar objectives, they differ in their approach to waste management: recycling focuses on transforming waste into new products, thereby reducing the amount sent to landfills; upcycling, instead, seeks to enhance the value of discarded materials by creating higher-value and more useful products without necessarily reprocessing them.

---

[8] https://www.food-hub.it/media/2022/09/16/fibre-tessili-scarti-alimentari-rifiuti-risorsa/.

[9] https://eur-lex.europa.eu/legal-content/IT.

Beyond recycling and upcycling, another promising option for sustainable textile production is the use of organic and food waste as raw materials for eco-friendly fiber production. Technological innovation has made it possible to transform food production waste such as fruit peels or corn leaves into new biodegradable textile materials [51], an innovative and promising approach to reducing the environmental impact of the textile industry while providing an alternative use for organic waste. Materials derived from biological sources also demonstrate versatility in their applications, ensuring high performance that meets consumer expectations.[10]

In this regard, an important initiative in sustainable fashion has been developed by the U.S.-based company Circular Systems,[11] a materials science enterprise committed to creating a positive impact on the environment, society, and economy through sustainable textile innovation. Its Agraloop Bio-Refinery[12] technology uses agricultural and post-consumer food waste as raw material to create natural fibers called BioFibre. Thanks to their plant-based chemistry, these wastes are transformed into textile fibers that can be used for clothing production. Circular Systems is one of several companies advancing research toward a more sustainable fashion industry by using science and technology to develop ecological and ethical textile materials. Other companies using innovative textile fibers include MycoWorks,[13] a U.S. company that uses fungal mycelium (i.e., the vegetative apparatus of mushrooms) to produce a sustainable, high-quality synthetic leather called *Reishi*; Ananas Anam,[14] a U.K. firm that produces eco-friendly leather for fashion and furniture from pineapple leaves; and Frumat,[15] an Italian company that transforms apple industry residues into biodegradable fabrics.

Such producers aim to use new sources of raw materials to create eco-friendly textiles that are as durable and high-quality as traditional ones. These innovations are driving a significant shift in the textile sector, fostering greater attention to sustainability and ecology, and promoting a greener future for the fashion industry.

Sustainable Textile Fibers

The premises discussed so far suggest that the textile industry has made significant progress due to growing awareness of the environmental impact of its activities. This awareness has led to the search for innovative and sustainable solutions for the production of textile fibers that minimize environmental impact while still meeting consumer demand [60].

According to the *Textile Exchange Market Report 2021*, global fiber production experienced a considerable decline, from 111 million tons in 2019 to 109 million tons in 2020, mainly due to the crisis caused by the COVID-19 pandemic. However, there

---

[10] https://www.fibre2fashion.com/industry-article/7858/transforming-the-fashion-industry-one-bio-based-fibre-at-a-time.

[11] https://circularsystems.com/about/.

[12] https://circularsystems.com/agraloop/.

[13] https://www.mycoworks.com/.

[14] https://www.ananas-anam.com/.

[15] https://www.today.it/speciale/game-changer/frumat-mele.html.

was an increase in the market share of eco-friendly or "green" fibers [61], confirming that they represent an important opportunity for promoting sustainable fashion.

As previously mentioned, this approach to fashion considers the environmental and social impact of the entire life cycle of a garment, from production to final distribution. Within this approach, the use of sustainable textile fibers can help reduce the ecological footprint of fashion and improve the overall sustainability of the textile industry [62].

Various types of sustainable textile fibers are available on the market. The most commonly discussed include:

- ***Recycled Fibers***, produced from waste materials such as plastic bottles or old fabrics, which are transformed into new textile fibers. This process reduces the amount of waste sent to landfills and contributes to the creation of a circular economy [63].
- ***Natural Fibers***, derived from natural resources such as animals or plants. They are not necessarily produced according to organic standards and are often cultivated using pesticides, chemical fertilizers, and other conventional agricultural practices. These include cotton, linen, hemp, silk, wool, and cashmere [64].
- ***Innovative Fibers***, developed through new technologies that require less water and energy for production compared to traditional fibers, either natural or recycled. Among these, fibers obtained from agricultural waste and other organic residues play a significant role, as they represent potential renewable and biodegradable materials for textile production [51].

Sustainable textile fibers can be used to create fabrics with specific properties such as moisture absorption, wrinkle resistance, breathability, and UV protection [65]. These properties can be achieved without the use of harmful chemicals, resulting in fabrics that are safer and "healthier" for consumers.

Overall, sustainable textile fibers represent an important opportunity for the textile industry to reduce its environmental impact and enhance sustainability in the fashion sector [66], by decreasing the use of natural resources, reducing CO_2 emissions, and improving textile waste management.

Recycled Fibers

Currently, in Europe, more than 15 kg of textile waste are generated per capita. The main source of this waste comes from discarded clothing, which accounts for about 85% of total textile waste.[16] This phenomenon represents a serious issue, as most textile waste ends up in landfills or is incinerated, producing negative effects on both the environment and human health [67].

To address this problem, several strategies have been proposed: reducing overproduction and overconsumption, designing longer-lasting materials, and creating products that promote circularity [68]. One of the most effective and sustainable solutions is fiber-to-fiber recycling, which allows textile waste to be transformed

[16] https://www.mckinsey.com/industries/retail/our-insights/scaling-textile-recycling-in-europe-turning-waste-into-value.

into usable fibers for creating new garments or other products [69]. Once recycled, these fibers represent a sustainable alternative to conventional ones, as they limit textile waste production and require fewer natural resources during processing. This is because they originate from post-consumer or post-industrial materials such as used clothing, fabric scraps, and manufacturing residues [70]. For these reasons, textile recycling is considered a key variable for achieving a sustainable transition in the fashion world, explaining why many fashion brands are now adopting it in their production processes [71].

For example, H&M began using recycled materials in 2009 and later launched its first *Conscious Collection* in 2010. The company is also recognized as the world's second-largest user of recycled polyester for textiles [71]. Another successful example in the sustainable fashion industry is Patagonia, which has successfully integrated textile recycling technologies into its product development process through its *WornWear Program* and *Common Threads Recycling Program.*

However, textile recycling remains a major challenge for many countries, with recycling rates in some cases still below 50%. Only a few countries, such as Canada, France, and Italy, show relatively higher recycling rates [72]. According to Leal Filho et al. [69], optimizing textile recycling is hindered by several factors, including economic feasibility, product composition, and the limited availability of recyclable textile materials.

Regarding economic feasibility, many waste materials are unsuitable for reuse or multiple cycles, mainly due to the widespread production of low-quality recycled products (*downcycling*), which discourages investment in textile recycling. As for textile composition, the structural elements of many fabrics make them incompatible with recycling processes, for example when they contain plastics or metallic residues. Furthermore, the quantity of textile waste suitable and available for recycling remains significantly insufficient.

Despite the numerous advantages of recycled fibers, such as reducing textile waste, lowering environmental impact, and saving natural resources, there are still some concerns about their quality and the potential environmental effects of recycling processes. In most cases, textile recycling requires mechanical, chemical, or biological processes that consume additional energy and raw materials, generating waste and potentially contaminating air, water, and soil [73].

These concerns also affect consumer perceptions and purchasing behavior. According to Polyportis et al. [74], although products made from recycled materials are considered eco-friendly and are often evaluated positively, many consumers still do not prefer them. This behavior stems from negative emotions associated with recycled products, particularly a perceived risk of contamination and an underlying feeling of disgust.

Another factor contributing to this reluctance is the perceived quality of recycled materials, often considered inferior to that of "original" materials. This perception reduces consumers' purchase intentions [75] and their willingness to pay for recycled products [76].

In conclusion, although recycled textile fibers hold great potential for contributing to the circular economy, many challenges must still be addressed to ensure their success and broader adoption.

Natural Fibers

A natural fiber is defined as any fiber that exists in its original state [64]. Such fibers have always played a crucial role in human history, providing materials for clothing, household textiles, and industrial applications. Unlike synthetic fibers, which are artificially produced from chemical compounds, natural fibers offer many advantages due to their abundance, easy availability, and low cost [77], as well as their sustainable and environmentally friendly properties [78]. For these reasons, demand for natural fibers has grown exponentially in recent decades [79]. In the textile sector, they can be divided into two main categories: *plant-based* and *animal-based fibers* [80].

Plant-Based Fibers

Plant-based fibers come from plants and are generally extracted from the stem, leaves, or seeds of the plant. They include:

- *Cotton*, extracted from the seeds of the cotton plant, is the most widely used natural fiber in the world. Soft, breathable, and durable, it is ideal for clothing and household textiles [81].
- *Linen*, obtained from the stems of the *Linum usitatissimum* plant, which has been cultivated and used for textile production since prehistoric times [82]. Linen possesses natural antibacterial properties and is appreciated for its strength, breathability, and thermoregulating capacity.
- *Hemp*, derived from the *Cannabis sativa* plant, is a strong, breathable, and biodegradable fiber with a low environmental impact, cultivated mainly in Europe and Asia [83].
- *Jute*, a biodegradable and eco-friendly fiber extracted from the jute plant, commonly used to make sacks, rugs, and ropes [84].

Animal-Based Fibers

Animal fibers are organic filaments derived from certain animals and used in the production of textiles and garments. The most common are:

- *Wool*, obtained mainly from sheep, is a natural insulating, breathable, and moisture-resistant fiber [85].
- *Silk*, produced by silkworms, is a highly valued fiber for its elegance, softness, and strength, suitable for high-end clothing and home furnishings. Silk fibers also have excellent mechanical properties, such as tensile strength, extensibility, and compressibility [86].
- *Cashmere*, obtained from cashmere goats, is a luxury fiber known for its sheen, elasticity, lightness, and softness [87].

These fibers offer a wide range of desirable properties that make them suitable for multiple applications. Being soft and breathable, most of them, ensure comfort and effective moisture management. Moreover, unlike artificial fibers, natural ones are biodegradable, meaning they decompose more quickly in the environment, reducing their ecological footprint and contributing to a more sustainable future [88].

However, some natural textiles still face issues related to their properties, such as low strength and limited durability [89], characteristics that may reduce their appeal to consumers. Furthermore, although natural textiles are generally perceived as ecologically sustainable, the intensification of agricultural and livestock practices associated with their production has generated negative effects on soil, sometimes causing severe and irreversible damage [90]. Natural fiber crops often require pesticides and chemical fertilizers, which can contaminate soil and groundwater and negatively affect the health of agricultural workers [91].

There are also ethical concerns related to the production of animal-derived natural fibers such as wool and silk. In commercial sheep farms, animals are often sheared hastily and carelessly, causing injuries [92]. Even more harmful for animal welfare is silk production, which typically involves boiling silkworms to extract the silk filament.

In light of these issues, to maximize the benefits of natural fibers while minimizing their negative impacts, it is essential to invest in technological innovation and increasingly sustainable and ethical production practices, considering both environmental and social aspects.

Innovative Fibers

The growing consumer interest in sustainable fashion is stimulating research into new materials and other forms of innovation in sustainable textile products, many of which fall under the category of environmental innovations [93]. In this context, *innovative textile fibers* refer to fibers whose production processes are novel and innovative, revolutionizing the textile world and offering new solutions for modern needs. These fibers can be derived from a wide variety of sources, including polymers, proteins, algae, fungi, and even food waste.

For instance, food waste contains a broad range of chemical compounds with great potential for exploitation across various industries. These compounds can be extracted, purified, and then used as raw materials for new products [94]. Thus, valorizing waste has considerable potential to provide economic, social, and environmental benefits. Several countries are already promoting strategies to support a circular economy in the agri-food sector by using waste as a resource [95]. Consequently, a growing number of companies are considering these innovations, leveraging opportunities provided by food and organic waste valorization.

One of the most promising innovations in this field is *mushroom-based fibers*, produced from mycelium (the vegetative part of fungi), which are transformed into a leather-like material [96]. The resulting fiber is soft, durable, and biodegradable, suitable for clothing, upholstery, and mattresses. These fibers are free from harmful chemicals and can be easily composted at the end of their lifecycle [97].

Another innovative fiber is produced by the Italian start-up *Orange Fiber*, which aims to create eco-friendly textiles from orange juice by-products [51]. The company developed and patented a nanotechnology-based process that replaces highly polluting raw materials with natural elements [53]. The resulting sustainable fabric has a silk-like texture, is soft to the touch, durable, breathable, and suitable for producing elegant, high-quality clothing.

Another example is the fiber derived from fermented dairy, known as QMilk (or Qmilch), which is obtained from whey, a byproduct of cheese production. The whey is fermented with specific bacteria, producing a silky, soft fiber resembling silk. This fiber is highly durable, breathable, and biodegradable, and can be used for a wide range of products: home textiles, bed linens, towels, everyday clothing, T-shirts, socks, and underwear. Even the automotive industry has shown interest in using this material for car seat production [98].

Among the best-known innovative sustainable fibers is Piñatex, a fabric made from pineapple leaves, which is produced using a low-impact process to create a strong and durable material [99]. Piñatex is a sustainable alternative to animal leather and is used to make bags, shoes, and other fashion accessories.

Marine algae-based fibers are also gaining popularity, such as SeaCell [100], a kelp-based textile developed by Italian designer Alberto Zanrè. This fiber is rich in antioxidant and remineralizing properties, making it soft and resistant to washing and ironing, similar to conventional cotton.

Innovative textile fibers represent an original and sustainable option for producing high-quality fabrics and accessories. Their biodegradable nature and eco-friendly properties offer solutions to textile waste and environmental pollution, providing materials that respect both the planet and human well-being.

However, using these innovative solutions often requires knowledge that companies may not possess internally. For this reason, brands are increasingly engaging in cooperative behaviors and adopting external sourcing strategies [101]. Many fashion brands now collaborate with external companies to access these sustainable fibers.

Recent collaborations include:

- Zara and Ananas Anam, which in 2022 created a collection of sneakers, sandals, and accessories made from Piñatex.
- Calvin Klein, which recently leveraged the expertise of Vegea, a low-impact leather company, to produce a line of men's sneakers, belts, and wallets.
- Stella McCartney, which featured Mylo ("unleather"), a mycelium-based material, in its Spring/Summer 2022 collection.

These collaborations allow fashion companies to access technologies and specialist knowledge that may not be available internally, while promoting innovation and sustainability in the sector [53]. They are enabling the creation of clothing and textiles made from innovative, eco-friendly materials that meet the growing demands of environmentally conscious consumers. The result is a transformation of the fashion industry toward more sustainable and responsible practices, reducing environmental impact and promoting a greener, sustainable future for all.

Fast Fashion and Luxury Fashion

Overall, the fashion apparel industry has evolved significantly over time, particularly in the past 20 years. The changing dynamics of the fashion industry, especially regarding consumers, have pushed retailers to demand low costs and high flexibility in design, quality, and speed to market, as well as the adoption of key strategies by companies to remain profitable in an increasingly demanding market [102].

Fast Fashion brands such as H&M and Zara emerged to meet a market characterized by speed, constant change, and high demand. They are defined as "*low-cost clothing collections that imitate luxury fashion trends*" [103], designed to bring the latest styles to consumers quickly so they can wear them during peak popularity and then discard them after a few uses. These brands share several key characteristics: rapid turnover, wide style variety, low labor costs, limited quantities to trigger impulse purchases, and the use of low-quality materials [87]. Building on these widely recognized attributes, [104] show that fast fashion consumption is socially interpreted as a signal of low self-control. These characteristics contribute to consumers' often ambivalent or negative implicit attitudes toward fast fashion. In this vein, [105] show that consumers' implicit attitudes and perceptions of warmth play a crucial role in shaping sustainable fashion consumption.

The "*Here Today, Gone Tomorrow*" logic [102] incentivizes frequent store visits and immediate, low-effort gratification. Despite their popularity, fast fashion brands face major sustainability challenges. Their business model generates high levels of waste: unsold and unworn items are discarded and often incinerated, and large-scale production contributes to significant environmental pollution. Chan et al. [106] show that some consumers recognize these environmental costs, experiencing guilt and cognitive dissonance between their desire to consume sustainably and their actual fast fashion purchases. Similarly, Mrad et al. [107] identify drivers of fast fashion consumption, including constant updates, perceived value, and large assortments, that reinforce dependency and consumption despite sustainability concerns.

Fast fashion brands have adopted various sustainable initiatives, but these often lead to contradictory reactions [108]. Consumers' perceptions of corporate social responsibility (CSR) influence attitudes toward brands [109], but CSR actions must be viewed as authentic and altruistic, with trust being fundamental. Yet the sector itself is inherently at odds with sustainability: speed, mass production, and rapid turnover clash with ecological values [110]. Cultural factors further complicate consumer responses. Kong et al. [111] show that sustainability resonates differently across countries, consumers from developed nations value sustainability more and are willing to pay a premium for green products. Fast fashion brands must therefore learn to associate their products with altruism, moderation, and ethics to satisfy environmentally conscious consumers.

As consumers have become more attentive to the social and environmental consequences of their purchases, luxury companies have increasingly recognized that they can no longer overlook sustainability, a topic they historically tended to ignore [112]. Luxury buyers expect "*credible responses to questions of environmental and social responsibility*" [113, p. 8], a shift partly driven by the broader transformation of the luxury sector in the last twenty years, namely, its democratization, which has made luxury accessible to the "*happy many*" and not solely to an elite minority [114, 115]. In parallel with this rising interest from both companies and consumers, academic research has increasingly contributing to a more optimistic perspective on sustainable luxury [110].

This view suggests that integrating sustainability into luxury offerings may enhance consumer evaluations and purchase intentions. One explanation is that luxury and sustainability share the notion of "rarity" [116, p. 41]. Luxury goods rely on scarce materials and exceptional craftsmanship [117], while sustainability emphasizes conserving limited resources and preventing overuse. Building on the idea of the "fallacy of clean luxury" [118, 119] argue that luxury brands may even be perceived as more sustainable than mass-market brands because of their emphasis on quality and prestige.

Empirical research provides additional nuance to this compatibility. For example, Janssen et al. [120] found that scarcity can increase the perceived social responsibility of durable luxury items (such as jewelry), enhancing attitudes toward the product, though this pattern does not extend to more ephemeral categories like fashion apparel. De Angelis et al. [121] further demonstrated that consumers respond more positively to luxury brands' eco-friendly offerings when these resemble products created by green, non-luxury firms rather than the brand's previous non-sustainable designs, suggesting that luxury can indeed be both "gold and green" (p. 1516). Other findings show that high-power consumers evaluate CSR initiatives more favorably [122], and that renting luxury apparel, a more sustainable form of consumption, may support status signaling [123], consistent with evidence that luxury buyers are highly motivated by self-presentation (e.g., [124]).

Sustainability is widely recognized not just as a moral or regulatory imperative, but also as a driver of innovation. Research shows that embedding environmental and social objectives into business strategy fosters novel products, processes, and business models, that is "sustainable innovation" [125]. In this light, the luxury fashion sector, known for its craftsmanship, heritage, and premium positioning, holds a distinctive opportunity to merge sustainability with innovation [126, 127]. Supply-chain innovation tailored to luxury fashion enables firms to redesign sourcing, logistics, and production systems while preserving the exclusivity and superior quality that characterize the industry [128, 129]. Moreover, adopting eco-innovation in materials, production technologies, or organizational practices aligns with broader trends in sustainable innovation and can strengthen a brand's competitive advantage without undermining its luxury essence [103, 130]. Consequently, rather than viewing luxury and sustainability as contradictory, luxury brands that strategically embrace innovation can harmonize exclusivity, high quality, and responsible practices, transforming sustainability challenges into creative opportunities and renewed brand value [117, 131].

2.2 Conceptual Model

Building on the growing relevance of sustainable materials in fashion consumption, it becomes essential to examine how consumers evaluate different types of eco-friendly textile fibers having a different *textile origin* within distinct brand contexts. Although sustainable production practices tend to enhance consumer attitudes and willingness to buy, not all sustainable materials carry the same symbolic and perceptual value. Research shows that materials perceived as *natural*, *organic*, or *biologically derived* foster stronger perceptions of purity, authenticity, and environmental care compared to materials associated with industrial processes such as plastic recycling [132, 133]. This distinction is particularly relevant in luxury fashion, a sector in which brand value is closely tied to symbolic attributes, sensory quality, and perceptions of craftsmanship and authenticity [131]. In this context, fibers derived from organic waste, such as Orange Fiber, may be perceived as more aligned with the values of naturalness, innovation, and refined sustainability that luxury brands increasingly seek to communicate, ultimately generating higher willingness to buy. Recycled plastic fibers, although sustainable, may instead evoke associations with waste, lower quality, or industrial processing, which conflict with luxury expectations [58]. By contrast, in fast fashion, where consumers anticipate lower durability, lower prices, and more standardized materials, the differentiation between organic waste-based fibers and recycled plastic fibers is likely attenuated, as the brand category itself sets lower expectations of sensory quality and symbolic value [118]. Therefore, the perceived fit between material type and brand positioning should shape consumers' responses to sustainable textile fibers. Consistent with this reasoning, we propose that the effect of textile origin (Orange Fiber versus recycled plastic fiber) on willingness to buy is moderated by brand type, such that the advantage of Orange Fiber over recycled plastic fiber is stronger for luxury brands and reduced for fast fashion brands. Formally:

H_1: The effect of the textile origin (Orange fiber versus Recycled plastic fiber) on willingness to buy is moderated by brand type (luxury versus fast fashion). Specifically, (a) for luxury brands, Orange fiber generates higher willingness to buy compared to Recycled plastic fiber, while (b) for fast fashion brands this effect is reduced.

We also propose that these effects are driven by brand coolness [3]. Indeed, when the material type is perceived as aligning with the brand's positioning, such perception of fit enhances the brand's perceived coolness, thus shaping consumers' responses to sustainable textile fibers [134]. Brand coolness refers to the extent to which a brand is perceived as autonomous, authentic, attractive, subcultural, original, and extraordinary [3]. As a multidimensional construct, brand coolness captures consumers' symbolic value and is especially sensitive to cues that signal innovation, authenticity, and value-congruent differentiation. Because of its strong identity-expressive function, it provides a particularly suitable lens for explaining how textile origin shapes consumers' willingness to buy across different brand types. The type of sustainable material incorporated into a product provides such cues and can generate symbolic value. Biologically derived fibers, such as Orange Fiber, convey stronger associations with purity, innovation, and sophisticated sustainability. These symbolic meanings can make the brand adopting such materials appear more authentic, trend-setting, and aesthetically appealing, thus cooler, compared to when the same brand employs recycled plastic fibers, which may evoke instead associations of waste, industrial processing, or lower sensory refinement. This effect is expected to be particularly pronounced for luxury brands, where perceptions of authenticity, craftsmanship, and symbolic distinctiveness represent core value drivers. For such brands, Orange Fiber is likely to be viewed as more consistent with brand values and therefore more capable of generating brand coolness [135]. Increased perceptions of coolness, in turn, enhance consumers' willingness to buy [136]. Conversely, for fast fashion brands, where expectations related to symbolic value, authenticity, and craftsmanship are intrinsically lower [137], the perceived coolness of Orange Fiber compared to recycled plastic is attenuated, leading to a weaker effect of textile origin (Orange Fiber versus recycled plastic) on willingness to buy. Formally:

H_2: The effect of the textile origin (Orange fiber versus Recycled plastic fiber) on willingness to buy, moderated by brand type (luxury versus fast fashion), is mediated by perceived coolness.

Figure 1 summarizes the conceptual model.

To test the research hypotheses, we conducted an experimental study concerning a travel bag made with eco-sustainable fibers, either innovative (i.e., Orange fiber) or traditional, (i.e., Recycled plastic fiber) produced by either a luxury fashion brand or a fast-fashion brand.

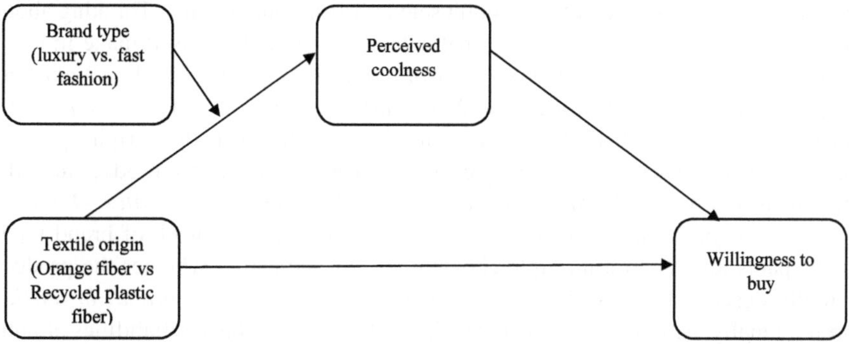

Fig. 1 Conceptual model. *Source* Authors' own elaboration

3 Empirical Evidence

The experimental study aimed to test H1 and H2. We used a 2 (Textile origin: Orange Fiber versus Recycled plastic fiber) by 2 (Brand type: luxury versus fast fashion) between-subjects design and measured brand coolness and willingness to pay for the product.

3.1 Method and Procedure

Participants were students and graduates from a large European university (N = 206; 52.4% female, 47.6% male; $M_{age} = 26.68$, $SD_{age} = 6.06$), who took part in an online study on a voluntary basis. Participants were informed that they would read a scenario describing either a luxury or fast fashion company that had recently launched a line of travel bags made from sustainable fibers and were randomly assigned to one of four experimental conditions. Depending on the assigned condition, participants were told that the company was either a luxury or fast fashion company, and that the travel bag was made with either an Orange Fiber or a recycled plastic fiber. The full scenarios are presented in Fig. 2.

After exposure to the scenario, participants indicated their willingness to buy for the product using three 7-point items: "The likelihood that you would purchase this product is:", "The likelihood that you would consider purchasing this product is:", "My willingness to purchase the product is:" (1 = *very low*, 7 = *very high*). Perceived coolness was assessed with four 7-point semantic differential items: Thinking about the travel bag described earlier, to what extent would the fiber used make it: (1 = *Not at all trendy/7* = *Very trendy*; 1 = *Not at all hip/7* = *Very hip*; 1 = *Not at all appealing/7* = *Very appealing*; 1 = *Not at all fascinating/7* = *Very fascinating*), adapted from Im et al. [138]. As a manipulation check of textile origin, participants answer to a multiple-choice question, indicating whether the product initially described was made with fibers: *Obtained from Recycled plastic bottles*, *Obtained from orange production*, *Classic textiles.* As a manipulation check of brand type, participants answer to a multiple-choice question, indicating whether the product initially described was: *a luxury fashion brand*, *a fast fashion brand*, *lower-tier brand*. Finally, participants indicated their gender and age. Since reliabilities of both willingness to buy ($\alpha = 0.95$) and perceived coolness ($\alpha = 0.95$) scales were high, we computed a composite mean score for subsequent analyses.

3.2 Results

The manipulation of the textile origin was successful. A Chi-square test revealed a strong and statistically significant association between the assigned fiber condition

		Brand type	
		Fast fashion	**Luxury**
Textile origin	**Recycled plastic fiber**	In the next screen, you will be presented with a scenario involving a fast fashion company that has recently launched a line of travel bags made with sustainable fibers. To avoid influencing your responses, the company will not be mentioned and will be kept anonymous (company X). The fast fashion company X has launched a line of travel bags made with fully sustainable fibers. Specifically, the fiber is derived from recycled plastic bottle waste that would otherwise be discarded and is instead given a completely new life. The sustainable fabric made from recycled plastic is described as comfortable, soft, and ecological, designed to meet the needs of a more sustainable fashion industry.	In the next screen, you will be presented with a scenario involving a fast fashion company that has recently launched a line of travel bags made with sustainable fibers. To avoid influencing your responses, the company will not be mentioned and will be kept anonymous (company X). The fast fashion company X has launched a line of travel bags made with fully sustainable fibers. Specifically, the fiber is derived from recycled plastic bottle waste that would otherwise be discarded and is instead given a completely new life. The sustainable fabric made from recycled plastic is described as comfortable, soft, and ecological, designed to meet the needs of a more sustainable fashion industry.
	Orange Fiber	In the next screen, you will be presented with a scenario involving a luxury fashion company that has recently launched a line of travel bags made with sustainable fibers. To avoid influencing your responses, the company will not be mentioned and will be kept anonymous (company X). The luxury company X has launched a line of travel bags made with fully eco-sustainable fibers. Specifically, the fiber is derived from orange production waste that would otherwise be discarded and is instead given a completely new life. The sustainable fabric made from orange fiber is described as exclusive, soft, and fully biodegradable, designed to meet the needs of a more sustainable fashion industry.	In the next screen, you will be presented with a scenario involving a luxury fashion company that has recently launched a line of travel bags made with sustainable fibers. To avoid influencing your responses, the company will not be mentioned and will be kept anonymous (company X). The luxury company X has launched a line of travel bags made with fully eco-sustainable fibers. Specifically, the fiber is derived from orange production waste that would otherwise be discarded and is instead given a completely new life. The sustainable fabric made from orange fiber is described as exclusive, soft, and fully biodegradable, designed to meet the needs of a more sustainable fashion industry.

Fig. 2 The four scenarios used in the study. *Source* Authors' own elaboration

and participants' responses to the manipulation check ($\chi^2(2) = 162.03, p < 0.001$. In the recycled-plastic condition, 91.6% of participants correctly identified the product as obtained from recycled plastic bottles, whereas 92.9% of participants in the orange-fiber condition correctly identified the product as obtained from orange production. Very few respondents in either condition chose '*Classic textiles*'. Cramer's V was $= 0.887, p < 0.001$), further confirming that participants accurately recognized the fiber type associated with their assigned condition.

The manipulation of the brand type was successful as well. A Chi-square test revealed a strong and statistically significant association between the assigned brand type condition and participants' responses to the manipulation check ($\chi^2(2) = 147.72$, $p < 0.001$). In the fast-fashion condition, 91.0% of participants correctly identified the product as produced by a fast-fashion brand, whereas 91.5% of participants in the luxury condition correctly identified the product as produced by a luxury brand. Very few respondents in either condition selected '*lower-tier brand*. Cramer's V was = 0.847, $p < 0.001$, further confirming that participants accurately recognized the brand type associated with their assigned condition.

A two-way ANOVA on willingness to buy score showed a significant main effect of textile origin ($F_{(1, 202)} = 50.313, p < 0.001$). Participants in the orange-fiber condition reported a higher willingness to buy ($M_{orange\text{-}fiber} = 4.98$, SD = 1.08) than participants in the recycled-plastic condition ($M_{recycled\text{-}plastic} = 3.84$, SD = 1.31). There was also a significant main effect of brand type ($F_{(1, 202)} = 13.173, p < 0.001$). Participants in the luxury brand condition reported a higher willingness to buy ($M_{luxury} = 4.65$, SD = 1.47) than participants in the fast-fashion brand condition ($M_{fast\ fashion} = 4.10$, SD = 1.09). More importantly, there was a significant textile origin × brand type interaction ($F_{(1, 202)} = 16.361, p < 0.001$). Planned comparisons revealed that, for a fast-fashion brand, an orange fiber origin produced higher willingness to buy than a recycled plastic origin ($M_{orange\text{-}fiber} = 4.35$, SD = 0.93; $M_{recycled\text{-}plastic} = 3.87$, SD = 1.19; $F_{(1, 202)} = 4.51, p = 0.035$). For a luxury brand, an orange fiber origin produced higher willingness to buy than a recycled plastic origin ($M_{orange\text{-}fiber} = 5.57$, SD = 0.85; $M_{recycled\text{-}plastic} = 3.81$, SD = 1.43; $F_{(1, 202)} = 63.89, p < 0.001$). Therefore, for a luxury brand, orange fiber origin produced higher willingness to buy than a recycled plastic origin, and this effect was stronger than a fast-fashion brand (Fig. 3). These results support H1.

To test the hypothesized mechanism underlying these effects, we estimated a moderated mediation model [139], in which the brand type (0 = fast fashion, 1 = luxury) moderates the path from the textile origin (0 = Recycled plastic fiber, 1 =

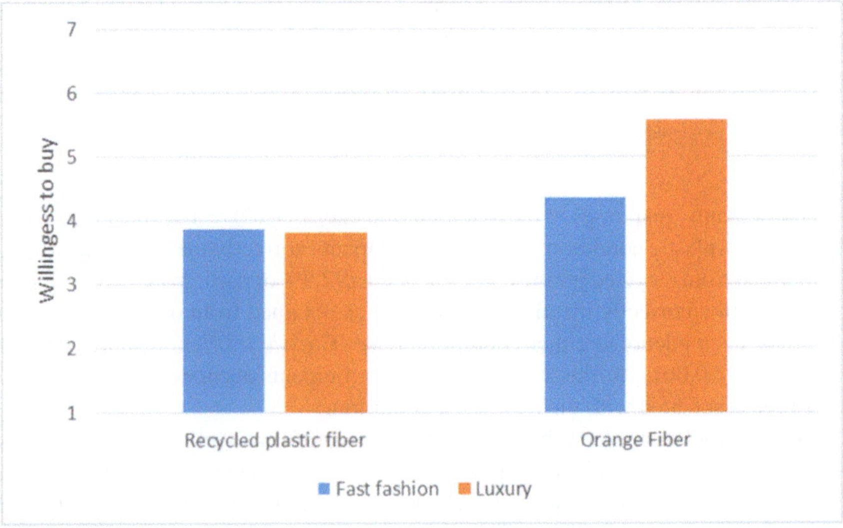

Fig. 3 The interaction between textile origin and brand type. *Source* Authors' own elaboration

Table 1 The indirect conditional effects

Conditional indirect effect	IE	BootLLCI	BootULCI
Fast fashion brands: textile origin → perceived coolness → willingness to buy	0.61	0.2585	1.0028
Luxury brands: textile origin → perceived coolness → willingness to buy	1.50	1.1626	1.8513

Source Authors' own elaboration

Orange Fiber) to perceived coolness; and in which perceived coolness influences willingness to buy.

Results showed that textile origin has a positive and significant effect on perceived coolness ($b = 0.81, p = 0.0003$), whereas the effect of brand type is not significant. More importantly, there was a significant textile origin × brand type on perceived coolness ($b = 1.18$, $p = 0.0001$). Simple slopes analysis showed that, for fashion brands, orange fibers made the product to be perceived as cooler than that made with recycled plastic ($b = 0.81, p = 0.0003$). For luxury brands, orange fibers still made the product to be perceived as cooler than that made with recycled plastic but with a greater intensity ($b = 1.99, p < 0.001$). In turn, perceived coolness positively influences willingness to buy ($b = 0.75$, $p < 0.001$), whereas when controlling for perceived coolness the effect of textile origin on willingness to buy is not significant. We found that, for fashion brands, the textile origin → perceived coolness → willingness to buy (IE = 0.61, 95% Bootstrap CI [0.2585, 1.0028]) conditional indirect effect was positive and significant. For luxury brands, the textile origin → perceived coolness → willingness to buy (IE = 1.50, 95% Bootstrap CI [1.1626, 1.8513]) conditional indirect effect was positive and significant (Table 1). The index of moderated mediation was significant (Index = 0.8867, 95% CI [0.4462, 1.3169]), indicating that the conditional indirect effects for the fast fashion and luxury brand conditions were significantly different from each other.

These results support H2 and provide evidence in support of the coolness-based mechanism driving the effects of the textile origin on willingness to buy for fast fashion and luxury brands.

4 Discussion

Although technical textiles for sustainability are gaining attention in the fashion industry, the comparative assessment of food-waste-derived fibers versus recycled plastics remains largely unexplored, especially in terms of their appeal to consumers across different brand types. Through an experimental study, we examined how textile origin (Orange Fiber versus recycled plastic fiber) influences consumer's willingness to buy, focusing on how these reactions vary according to brand positioning (luxury versus fast fashion). We showed that Orange Fiber generates higher willingness to buy than recycled plastic fiber when associated with luxury brands. The advantage of Orange Fiber over recycled plastic fiber is attenuated for fast fashion brands. The effect of textile origin on willingness to buy, moderated by the brand type, is mediated by perceived coolness [3]. Indeed, Orange Fiber conveys symbolic meanings of purity, innovation, and sophisticated sustainability, which enhance perceived coolness and, in turn, willingness to buy, particularly for luxury brands, where these values align with brand identity. For fast fashion brands, this effect is weaker due to lower expectations of symbolic value and craftsmanship that attenuate perceived coolness and, in turn, willingness to buy. These findings suggest that innovative technical textiles for sustainability combined with strong brand positioning can drive consumer adoption by joining sustainability and desirability.

Implications for Theory

This study makes three main contributions. First, we contribute to the literature on technical textiles for sustainability in fashion by demonstrating that not all sustainable materials are evaluated equally by consumers. Our study showing that Orange Fiber generates higher willingness to buy than Recycled plastic fiber, particularly in luxury contexts, extends sustainability research that has highlighted the growing relevance of innovative bio-based and upcycled materials [58, 64, 88]. By showing that perceived coolness mediates this effect, the study advances theories of sustainable consumption that increasingly emphasize aesthetic, symbolic, and identity-driven mechanisms alongside environmental or moral drivers [110, 112].

Second, this research contributes to the literature comparing luxury and fast fashion brands by showing that brand type fundamentally shapes the meaning and impact of sustainable cues. In luxury contexts, the use of Orange Fiber increases willingness to buy because sustainability, when supported by innovative materials and production processes, enhances symbolic value, status signaling, and stylistic differentiation. This pattern aligns with prior research demonstrating that luxury and sustainability can coexist when sustainability acquires a distinctive, identity-relevant meaning [58, 110]. Sustainable luxury products tend to be perceived as atypical and thus capable of reinforcing the exclusivity and uniqueness central to luxury consumption. Moreover, the integration of sustainability often amplifies the social and symbolic benefits that consumers derive from luxury brands, consistent with evidence showing that Corporate Social Responsibility initiatives in luxury contexts activate status-oriented and conspicuous motives [117]. In contrast, fast fashion operates within a schema dominated by affordability, immediacy, and rapid trend turnover. Within this meaning system, the origin of the textile represents a weaker cue, and its symbolic or identity-expressive potential is substantially diluted. The reduced effect of Orange Fiber in fast fashion contexts, therefore, refines comparative frameworks by demonstrating that sustainability does not act as a universal driver. Instead, its interpretive and motivational significance emerges through its interaction with brand positioning, activating different psychological pathways across luxury and fast fashion categories.

Finally, we contribute to the literature on brand coolness by extending its understanding as a key mechanism shaping consumer responses to sustainable materials in the fashion industry. Specifically, our findings advance insights into the drivers of brand coolness by demonstrating that sustainable material choices convey symbolic value, in addition to their environmental relevance, which enhances a brand's perceived coolness [3, 134]. Moreover, we highlight the role of brand context, showing that the effect of coolness is contingent on brand type and emphasizing that fit between brand values and the symbolic meanings of materials is crucial for generating perceived coolness and driving consumer adoption [135, 137].

Implication for Practice

The findings offer actionable guidance for managers in both luxury and fast fashion segments seeking to integrate technical textiles for sustainability into their offerings. First, luxury fashion brands benefit from emphasizing innovative sustainable materials such as Orange Fiber. Consumers in luxury contexts prefer bio-based and technologically advanced materials over recycled plastics, as these options enhance perceived coolness, symbolic value, and status signaling. Managers should therefore invest in innovative sustainable materials and leverage *coolness* as a central advertising and positioning strategy. Highlighting innovation, exclusivity, and aesthetic appeal can strengthen the brand's luxury identity while promoting environmental responsibility.

Second, fast fashion brands can also incorporate sustainability in their strategies, but the approach should differ. In this context, mainstream solutions such as recycled plastics are more suitable, aligning with consumers' focus on affordability, trendiness, and rapid product turnover. Although recycled plastics may evoke contamination concerns in luxury segments, they are a viable option in fast fashion, where symbolic meaning is less critical and functional or environmental benefits are prioritized. Managers can implement green initiatives using familiar, cost-effective materials that resonate with mass-market consumers.

Finally, given the coolness-based mechanism driving the proposed effect, managers should ensure that the use of innovative fibers is communicated in ways that highlight their symbolic and aesthetic value, particularly for luxury brands aiming to position themselves as trend-setting, authentic, and innovative. Marketing communications, product storytelling, and in-store experiences should emphasize how these materials reinforce the brand's identity, leveraging the symbolic cues embedded in sustainable innovations.

Limitations and Future Research

This study has some limitations that suggest avenues for future research. First, our focus was on the younger generation, leaving open the question of how older consumers perceive and accept innovative sustainable fibers compared to more traditional ones. Future studies could explore older age groups to examine potential generational differences in attitudes toward sustainability. Second, we employed anonymous brands in our experimental study. While this approach allowed us to isolate the effects of the manipulated variables, using real brands in future research could provide insights into how brand familiarity and reputation interact with perceptions of sustainable fibers. Third, we relied on a reduced, pre-validated scale to measure brand coolness [138]. Future research could employ the full, multidimensional version of the scale [3] to identify which specific dimensions of coolness are most strongly elicited by sustainable fibers. Fourth, our study used only textual scenarios without any visual representations. Future research could incorporate real product images or physical samples to enhance ecological validity. Storytelling could also be tested as a strategy to promote sustainable fibers, allowing researchers to explore which types of narratives generate stronger effects. Finally, alternative explanations for our findings cannot be fully ruled out. For example, perceptions of innovativeness or overall product quality may also influence consumer responses. Additionally, other innovative sustainable fibers could be studied to broaden understanding and guide product development.

This research highlights the importance of understanding how sustainability, innovation, and the emerging role of technical textiles influence purchasing decisions among key consumer segments in the fashion industry. Overall, the chapter emphasizes that innovative sustainable textiles, particularly technical textiles derived from food waste, such as orange fiber, represent not only an ecological solution but also a strategic opportunity for brands to strengthen market differentiation, appeal, and ethical positioning in the competitive fashion landscape.

References

1. Koller M, Floh A, Zauner A (2011) Further insights into perceived value and consumer loyalty: a "green" perspective. Psychol Market 28(12):1154–1176
2. Schiaroli V, Fraccascia L, Dangelico RM (2024) How can consumers behave sustainably in the fashion industry? A systematic literature review of determinants, drivers, and barriers across the consumption phases. J Clean Prod 483:144232
3. Warren C, Batra R, Loureiro SMC, Bagozzi RP (2019) Brand coolness. J Mark 83(5):36–56
4. Chen TB, Chai LT (2010) Attitude towards the environment and green products: consumers' perspective. Manag Sci Eng 4(2):27–39
5. World Commission on Environment and Development (1987) Our common future. Oxford University Press
6. Veleva V, Ellenbecker M (2001) Indicators of sustainable production: framework and methodology. J Clean Prod 9(6):519–549
7. Moisander J (2007) Motivational complexity of green consumerism. Int J Consum Stud 31(4):404–409
8. Ofstad S, Westly L, Bratelli T, Miljoverndepartementet N (1994) Symposium: sustainable consumption: 19–20 January 1994: Oslo. Ministry of Environment, Oslo, Norway
9. Testa F, Pretner G, Iovino R, Bianchi G, Tessitore S, Iraldo F (2021) Drivers to green consumption: a systematic review. Environ Dev Sustain 23:4826–4880
10. Abbaspour M, Aflaki E, Nejad FM (2019) Reuse of waste tire textile fibers as soil reinforcement. J Clean Prod 207:1059–1071
11. Zelezny LC, Schultz P (2000) Psychology of promoting environmentalism: promoting environmentalism. J Soc Issues 56(3):365–371
12. Da Giau A, Macchion L, Caniato F, Caridi M, Danese P, Rinaldi R, Vinelli A (2016) Sustainability practices and web-based communication: an analysis of the Italian fashion industry. J Fash Mark Manag 20(1):72–88
13. Donthu N, Gustafsson A (2020) Effects of COVID-19 on business and research. J Bus Res 117:284–289

14. Pantano E, Pizzi G, Scarpi D, Dennis C (2020) Competing during a pandemic? Retailers' ups and downs during the COVID-19 outbreak. J Bus Res 116:209–213
15. Degli Esposti P, Mortara A, Roberti G (2021) Sharing and sustainable consumption in the era of COVID-19. Sustainability 13(4):1903
16. Griskevicius V, Tybur JM, Van den Bergh B (2010) Going green to be seen: status, reputation, and conspicuous conservation. J Pers Soc Psychol 98(3):392
17. López-Pérez ME, Melero I, Javier Sese F (2017) Management for sustainable development and its impact on firm value in the SME context: does size matter? Bus Strateg Environ 26(7):985–999
18. Aguilera-Caracuel J, Guerrero-Villegas J (2018) How corporate social responsibility helps MNEs to improve their reputation: the moderating effects of geographical diversification and operating in developing regions. Corp Soc Responsib Environ Manag 25(4):355–372
19. Schmidt FC, Zanini RR, Korzenowski AL, Schmidt R, Nascimento KBX (2018) Evaluation of sustainability practices in small and medium-sized manufacturing enterprises in Southern Brazil. Sustainability 10(7):2460
20. Farzin M, Shababi H, Shirchi Sasi G, Sadeghi M, Makvandi R (2023) The determinants of eco-fashion purchase intention and willingness to pay. Spanish J Market ESIC 27(3):348–366
21. Delmas MA, Burbano VC (2011) The drivers of greenwashing. Calif Manage Rev 54(1):64–87
22. Chen YS, Chang CH (2013) Greenwash and green trust: the mediation effects of green consumer confusion and green perceived risk. J Bus Ethics 114(3):489–500
23. Rahman I, Park J, Chi C (2015) Consequences of 'greenwashing.' Int J Contemp Hosp Manag 27(6):1054–1081
24. Aji HM, Sutikno B (2015) The extended consequence of greenwashing: Perceived consumer skepticism. Int J Bus Inform 10(4):433
25. Epstein MJ, Roy MJ (2003) Making the business case for sustainability: linking social and environmental actions to financial performance. J Corp Citizsh 9:79–96
26. Ozdamar Ertekin Z, Atik D (2015) Sustainable markets: motivating factors, barriers, and remedies for mobilization of slow fashion. J Macromark 35(1):53–69
27. Cervellon MC, Hjerth H, Ricard S, Carey L (2010) Green in fashion? An exploratory study of national differences in consumers concern for eco-fashion. In: Proceedings of the 9th international marketing trends conference
28. McNeill LS, Snowdon J (2019) Slow fashion: balancing the conscious retail model within the fashion marketplace. Austr Mark J 27(4):215–223
29. Gazzola P, Pavione E, Pezzetti R, Grechi D (2020) Trends in the fashion industry. The perception of sustainability and circular economy: a gender/generation quantitative approach. Sustainability 12(7):2809
30. Shen B, Wang Y, Lo CK, Shum M (2012) The impact of ethical fashion on consumer purchase behavior. J Fash Market Manag Int J 16(2):234–245
31. Maheshwari V, Fielding J (2018) Importance of ethics and sustainability in the fashion industry: an abstract. Academy of marketing science world marketing congress. Springer, Cham, pp 139–140
32. Joergens C (2006) Ethical fashion: myth or future trend? J Fash Market Manag Int J 10(3):360–371
33. Carey L, Cervellon M-C (2014) Ethical fashion dimensions: pictorial and auditory depictions through three cultural perspectives. J Fash Mark Manag 18(4):483–506
34. Jung S, Jin B (2014) A theoretical investigation of slow fashion: sustainable future of the apparel industry. Int J Consum Stud 38(5):510–519
35. Wadhawan D, Ansari MF (2018) Sustainable fashion in India
36. Fletcher K (2010) Slow fashion: an invitation for systems change. Fash Pract J Des Creat Process Fash 2(2):259–266
37. Islam MM, Perry P, Gill S (2021) Mapping environmentally sustainable practices in textiles, apparel and fashion industries: a systematic literature review. J Fash Market Manag Int J 25(2):331–353

38. Hill J, Lee HH (2012) Young generation Y consumers' perceptions of sustainability in the apparel industry. J Fash Market Manag Int J 16(4):477–491
39. Blättel-Mink B (1998) Innovation towards sustainable economy-the integration of economy and ecology in companies. Sustain Dev 6(2):49–58
40. Mirata M, Emtairah T (2005) Industrial symbiosis networks and the contribution to environmental innovation: the case of the Landskrona industrial symbiosis programme. J Clean Prod 13(10–11):993–1002
41. Boons F, Lüdeke-Freund F (2013) Business models for sustainable innovation: state-of-the-art and steps towards a research agenda. J Clean Prod 45:9–19
42. Martínez P, del Bosque IR (2014) Sustainability dimensions: a source to enhance corporate reputation. Corp Reput Rev 17:239–253
43. Pereira Á, Vence X (2012) Key business factors for eco-innovation: an overview of recent firm-level empirical studies. Cuadernos de Gestion 12:73–103
44. Geisendorf S, Pietrulla F (2018) The circular economy and circular economic concepts—a literature analysis and redefinition. Thunderbird Int Bus Rev 60(5):771–782
45. Rinaldi FR (2021) Moda sostenibile e circolare tra ambiente, etica e cultura del territorio. Dimensioni e Problemi Della Ricerca Storica 16(2):197–214
46. Dissanayake DGK, Weerasinghe D (2022) Towards circular economy in fashion: review of strategies, barriers and enablers. Circ Econ Sustain 2(1):25–45
47. Yan RN, Hyllegard KH, Blaesi LF (2012) Marketing eco-fashion: the influence of brand name and message explicitness. J Mark Commun 18(2):151–168
48. Niinimäki K, Peters G, Dahlbo H, Perry P, Rissanen T, Gwilt A (2020) The environmental price of fast fashion. Nat Rev Earth Environ 1(4):189–200
49. Lee MS, Seifert M, Cherrier H (2017) Anti-consumption and governance in the global fashion industry: transparency is key. Governing corporate social responsibility in the apparel industry after Rana Plaza. Palgrave Macmillan US, New York, pp 147–174
50. Niinimäki K (2017) Fashion in a circular economy. Sustainability in fashion: a cradle to upcycle approach. Springer, Cham, pp 151–169
51. Sachidhanandham A (2020) Textiles from orange peel waste. J Sci Technol Develop 23(2):508–516
52. Ribul M, Lanot A, Pisapia CT, Purnell P, McQueen-Mason SJ, Baurley S (2021) Mechanical, chemical, biological: moving towards closed-loop bio-based recycling in a circular economy of sustainable textiles. J Clean Prod 326:129325
53. Todeschini BV, Cortimiglia MN, Callegaro-de-Menezes D, Ghezzi A (2017) Innovative and sustainable business models in the fashion industry: entrepreneurial drivers, opportunities, and challenges. Bus Horiz 60(6):759–770
54. Sandin G, Roos S, Johansson M (2019) Environmental impact of textile fibers: what we know and what we don't know: fiber Bible part 2
55. Skogen K, Helland H, Kaltenborn B (2018) Concern about climate change, biodiversity loss, habitat degradation and landscape change: embedded in different packages of environmental concern? J Nat Conserv 44:12–20
56. Damayanti D, Wulandari LA, Bagaskoro A, Rianjanu A, Wu HS (2021) Possibility routes for textile recycling technology. Polymers 13(21):3834
57. Heikkilä P, Cura K, Heikkilä J, Hinkka V, Ikonen T, Kamppuri T et al (2019) Telaketju: towards circularity of textiles
58. Adıgüzel F, Donato C (2021) Proud to be sustainable: Upcycled versus recycled luxury products. J Bus Res 130:137–146
59. Bridgens B, Powell M, Farmer G, Walsh C, Reed E, Royapoor M, Heidrich O (2018) Creative upcycling: reconnecting people, materials and place through making. J Clean Prod 189:145–154
60. Girijappa YGT, Rangappa SM, Parameswaranpillai J, Siengchin S (2019) Natural fibers as sustainable and renewable resource for development of eco-friendly composites: a comprehensive review. Front Mater 6:226

61. Kim T, Kim D, Park Y (2022) Recent progress in regenerated fibers for "green" textile products. J Clean Prod 12:134226
62. Sigaard AS, Laitala K (2023) Natural and sustainable? Consumers' textile fiber preferences. Fibers 11(2):12
63. Leonas KK (2017) The use of recycled fibers in fashion and home products. Text Cloth Sustain Recyc Upcyc Text Fash 12:55–77
64. da Silva CJG, de Medeiros ADLM, de Amorim JDP, Nascimento HA, Converti A, Costa AFS, Sarubbo LA (2021) Bacterial cellulose biotextiles for the future of sustainable fashion: a review. Environ Chem Lett 19(4):2967–2980
65. Shah JN, Padhye R, Pachauri RD (2022) Studies on UV protection and antimicrobial functionality of textiles. J Nat Fibers 19(13):6810–6821
66. Muthu SS, Gardetti MA (eds) (2016) Sustainable fibres for fashion industry, vol 1. Springer, Singapore
67. Zhou Q, Le QV, Meng L, Yang H, Gu H, Yang Y et al (2022) Environmental perspectives of textile waste, environmental pollution and recycling. Environ Technol Rev 11(1):62–71
68. Todor MP, Bulei C, Heput T, Kiss I (2018) Researches on the development of new composite materials complete/partially biodegradable using natural textile fibers of new vegetable origin and those recovered from textile waste. In: IOP conference series: materials science and engineering, vol 294. IOP Publishing, p 012021
69. Leal Filho W, Ellams D, Han S, Tyler D, Boiten VJ, Paço A et al (2019) A review of the socio-economic advantages of textile recycling. J Clean Prod 218:10–20
70. Dahlbo H, Aalto K, Eskelinen H, Salmenperä H (2017) Increasing textile circulation—consequences and requirements. Sustain Prod Consump 9:44–57
71. Watson D, Gylling AC, Andersson T, Heikkilä P (2017) Textile-to-textile recycling: ten Nordic brands that are leading the way
72. Conke LS (2018) Barriers to waste recycling development: evidence from Brazil. Resour Conserv Recycl 134:129–135
73. Zhou Y, Wu W, Qiu K (2010) Recovery of materials from waste printed circuit boards by vacuum pyrolysis and vacuum centrifugal separation. Waste Manage 30(11):2299–2304
74. Polyportis A, Mugge R, Magnier L (2022) Consumer acceptance of products made from recycled materials: a scoping review. Resour Conserv Recycl 186:106533
75. Queiroz FCBP, Lima NC, da Silva CL, Queiroz JV, de Souza GHS (2021) Purchase intentions for Brazilian recycled PET products: circular economy opportunities. Recycling 6(4):75
76. Hamzaoui-Essoussi L, Linton JD (2014) Offering branded remanufactured/recycled products: at what price? J Remanufact 4:1–15
77. Arpitha GR, Sanjay MR, Senthamaraikannan P, Barile C, Yogesha B (2017) Hybridization effect of sisal/glass/epoxy/filler based woven fabric reinforced composites. Exp Tech 41(6):577–584
78. Sajin JB, Paul RC, Binoj JS, Mansingh BB, Selvan MGA, Goh KL et al (2022) Impact of fiber length on mechanical, morphological and thermal analysis of chemical treated jute fiber polymer composites for sustainable applications. Curr Res Green Sustain Chem 5:100241
79. Akatwijuka O, Gepreel MAH, Abdel-Mawgood A, Yamamoto M, Saito Y, Hassanin AH (2022) Overview of banana cellulosic fibers: agro-biomass potential, fiber extraction, properties, and sustainable applications. Biomass Conver Bioref 16:1–17
80. Nayak R, Houshyar S, Khandual A, Padhye R, Fergusson S (2020) Identification of natural textile fibres. In: Handbook of natural fibres. Woodhead Publishing, pp 503–534
81. Elmogahzy Y, Farag R (2018) Tensile properties of cotton fibers: importance, research, and limitations. In: Handbook of properties of textile and technical fibres. Woodhead Publishing, pp 223–273
82. Bourmaud A, Siniscalco D, Foucat L, Goudenhooft C, Falourd X, Pontoire B et al (2019) Evolution of flax cell wall ultrastructure and mechanical properties during the retting step. Carbohyd Polym 206:48–56
83. Bhoopathi R, Ramesh M, Deepa C (2014) Fabrication and property evaluation of banana-hemp-glass fiber reinforced composites. Proced Eng 97:2032–2041

84. Khan JA, Khan MA (2015) The use of jute fibers as reinforcements in composites. In: Biofiber reinforcements in composite materials. Woodhead Publishing, pp 3–34
85. Allafi FAS, Hossain MS, Ab Kadir MO, Shaah MAH, Lalung J, Ahmad MI (2021) Waterless processing of sheep wool fiber in textile industry with supercritical CO_2: potential and challenges. J Clean Prod 285:124819
86. Yuan Q, Yao J, Chen X, Huang L, Shao Z (2010) The preparation of high performance silk fiber/fibroin composite. Polymer 51(21):4843–4849
87. Sun Y, Cai HH, Su R, Shen Q (2020) Advantage of low quality in short life cycle products. Asia Pac J Mark Logist 32(5):1038–1054
88. Costa AFS, Rocha MAV, Sarubbo LA (2017) Bacterial cellulose: an ecofriendly biotextile. Int J Text Fash Technol 7:11–26
89. Islam JM, Mondal MIH, Das SC (2021) The life and durability issues of natural textiles and clothing. In: Fundamentals of natural fibres and textiles. Woodhead Publishing, pp 657–690
90. Karthik T, Gopalakrishnan D (2014) Environmental analysis of textile value chain: an overview. In: Roadmap to sustainable textiles and clothing: environmental and social aspects of textiles and clothing supply chain, pp 153–188
91. Chen HL, Burns LD (2006) Environmental analysis of textile products. Cloth Text Res J 24(3):248–261
92. Lee C, Fisher AD (2007) Welfare consequences of mulesing of sheep. Aust Vet J 85(3):89–93
93. Kemp R, Arundel A, Smith K (1998) Survey indicators for environmental innovation. IDEA Paper Series 8:1998
94. Garcia-Garcia G, Stone J, Rahimifard S (2019) Opportunities for waste valorisation in the food industry: a case study with four UK food manufacturers. J Clean Prod 211:1339–1356
95. Fisgativa H, Tremier A, Le Roux S, Bureau C, Dabert P (2017) Understanding the anaerobic biodegradability of food waste: relationship between the typological, biochemical and microbial characteristics. J Environ Manage 188:95–107
96. Ekegårdh N, Kristersson E (2021) From fungi to vegan leather: a case study made to highlight the potentials of mycelium-based leatherin the textile industry and the unique features of its value chain
97. Vandelook S, Elsacker E, Van Wylick A, De Laet L, Peeters E (2021) Current state and future prospects of pure mycelium materials. Fung Biol Biotechnol 8(1):1–10
98. Grancarić AM, Jerković I, Tarbuk A (2013) Bioplastics in textiles. Polimeri Časopis za Plastiku i Gumu 34(1):9–14
99. Meyer M, Dietrich S, Schulz H, Mondschein A (2021) Comparison of the technical performance of leather, artificial leather, and trendy alternatives. Coatings 11(2):226
100. Cenkkut-Gültekin B (2016) Bleaching of SeaCell® active fabrics with hydrogen peroxide. Fibers Polym 17:1175–1180
101. De Marchi V (2012) Environmental innovation and R&D cooperation: empirical evidence from Spanish manufacturing firms. Res Policy 41(3):614–623
102. Bhardwaj V, Fairhurst A (2010) Fast fashion: response to changes in the fashion industry. Int Rev Retail Distrib Consum Res 20(1):165–173
103. Joy A, Sherry JF, Venkatesh A, Wang J, Chan R (2012) Fast fashion, sustainability, and the ethical appeal of luxury brands. Fash Theory 16(3):273–295
104. Huang Y, Zhang K, Deng X, Zhang Q (2025) Fast fashion consumption signals low self-control. J Consum Res 6:ucaf032
105. Grazzini L, Acuti D, Aiello G (2021) Solving the puzzle of sustainable fashion consumption: the role of consumers' implicit attitudes and perceived warmth. J Clean Prod 287:125579
106. Chan HK, He H, Wang WY (2012) Green marketing and its impact on supply chain management in industrial markets. Ind Mark Manage 41(4):557–562
107. Mrad M, Majdalani J, Cui CC, El Khansa Z (2020) Brand addiction in the contexts of luxury and fast-fashion brands. J Retail Consum Serv 55:102089
108. McNeill L, Moore R (2015) Sustainable fashion consumption and the fast fashion conundrum: fashionable consumers and attitudes to sustainability in clothing choice. Int J Consum Stud 39(3):212–222

109. Neumann HL, Martinez LM, Martinez LF (2021) Sustainability efforts in the fast fashion industry: consumer perception, trust and purchase intention. Sustain Account Manag Policy J 12(3):571–590
110. Amatulli C, De Angelis M, Donato C (2021) The atypicality of sustainable luxury products. Psychol Mark 38(11):1990–2005
111. Kong HM, Witmaier A, Ko E (2021) Sustainability and social media communication: how consumers respond to marketing efforts of luxury and non-luxury fashion brands. J Bus Res 131:640–651
112. Athwal N, Wells VK, Carrigan M, Henninger CE (2019) Sustainable luxury marketing: a synthesis and research agenda. Int J Manag Rev 21:405–426
113. Bendell J, Kleanthous A (2007) Deeper luxury: quality and style when the world matters. WWF-UK
114. Dubois B, Laurent G (1996) The functions of luxury: a situational approach to excursionism. ACR North Am Adv 2:452
115. Moraes C, Carrigan M, Bosangit C, Ferreira C, McGrath M (2017) Understanding ethical luxury consumption through practice theories: a study of fine jewellery purchases. J Bus Ethics 145(3):525–543
116. Kapferer J-N (2010) All that glitters is not green: the challenge of sustainable luxury. Eur Bus Rev 14:40–45
117. Amatulli C, De Angelis M, Korschun D, Romani S (2018) Consumers' perceptions of luxury brands' CSR initiatives: an investigation of the role of status and conspicuous consumption. J Clean Prod 194:277–287
118. Davies IA, Lee Z, Ahonkhai I (2012) Do consumers care about ethical-luxury? J Bus Ethics 106(1):37–51
119. Heil OP, Langer DA (2017) Identifying the luxury sustainability paradox: three steps toward a solution. In: Gardetti M (ed) Sustainable management of luxury: environmental footprints and eco-design of products and processes. Springer, pp 125–144
120. Janssen C, Vanhamme J, Lindgreen A, Lefebvre C (2014) The Catch-22 of responsible luxury: effects of luxury product characteristics on consumers' perception of fit with corporate social responsibility. J Bus Ethics 119(1):45–57
121. De Angelis M, Adıgüzel F, Amatulli C (2017) The role of design similarity in consumers' evaluation of new green products: an investigation of luxury fashion brands. J Clean Prod 141:1515–1527
122. Chang DR, Jang J, Lee H, Nam M (2019) The effects of power on consumers' evaluation of a luxury brand's corporate social responsibility. Psychol Mark 36(1):72–83
123. Pantano E, Stylos N (2020) The Cinderella moment: exploring consumers' motivations to engage with renting as collaborative luxury consumption mode. Psychol Market 37(5):740–753
124. Fazeli Z, Shukla P, Perks K (2020) Digital buying behavior: the role of regulatory fit and self-construal in online luxury goods purchase intentions. Psychol Mark 37(1):15–26
125. Dangelico RM, Pontrandolfo P, Pujari D (2013) Developing sustainable new products in the textile and upholstered furniture industries: role of external integrative capabilities. J Prod Innov Manag 30(4):642–658
126. Brun A, Castelli C (2013) The nature of luxury: a consumer perspective. Int J Retail Distrib Manag 41(11/12):823–847
127. Kapferer JN, Bastien V (2012) Luxe oblige. Editions Eyrolles
128. Brun A, Caniato F, Caridi M, Castelli C, Miragliotta G, Ronchi S et al (2008) Logistics and supply chain management in luxury fashion retail: empirical investigation of Italian firms. Int J Prod Econ 114(2):554–570
129. Caniato F, Moretto A, Caridi M (2013) Dynamic capabilities for fashion-luxury supply chain innovation. Int J Retail Distrib Manag 41(11/12):940–960
130. Niinimäki K (2013) Sustainable fashion: new approaches. Aalto University
131. Kapferer JN, Michaut-Denizeau A (2017) Is luxury compatible with sustainability? Luxury consumers' viewpoint. Advances in luxury brand management. Springer, Cham, pp 123–156

132. Magnier L, Crié D (2015) Communicating packaging eco-friendliness: an exploration of consumers' perceptions of eco-designed packaging. Int J Retail Distrib Manag 43(4–5):350–366
133. Steinhart Y, Ayalon O, Puterman H (2013) The effect of an environmental claim on consumers' perceptions about luxury and utilitarian products. J Clean Prod 53:277–286
134. Guerreiro J, Loureiro SMC, Nascimento J, Duarte M (2023) How to earn a premium price: the effect of green marketing and brand coolness. J Commun Manag 27(1):35–63
135. Loureiro SMC, Jiménez-Barreto J, Romero J (2020) Enhancing brand coolness through perceived luxury values: insight from luxury fashion brands. J Retail Consum Serv 57:102–211
136. Bagozzi RP, Khoshnevis M (2023) How and when brand coolness transforms product quality judgments into positive word of mouth and intentions to buy/use. J Market Theory Pract 31(4):383–402
137. Loureiro SMC, Breazeale M (2024) Luxury fashion brand coolness: niche versus mass cool. Spanish J Market 28(2):165–186
138. Im S, Bhat S, Lee Y (2015) Consumer perceptions of product creativity, coolness, value and attitude. J Bus Res 68(1):166–172
139. Hayes AF (2022) Introduction to mediation, moderation, and conditional process analysis: a regression-based approach, 3rd edn. The Guilford Press, New York

Biodegradability of Technical Textiles—A Review

Md. Shah Ikbal, Tohidul Kaiser, and Mohammad Abbas Uddin

Abstract Technical textiles constitute a rapidly expanding, high-value segment with diverse applications in healthcare, construction, automotive manufacturing, protective equipment, sports, agriculture, and packaging. These materials are engineered to provide specialized characteristics such as enhanced strength, durability, flexibility, chemical resistance, flame retardancy, and advanced functional coatings, which are often challenging to obtain through conventional means. While these properties deliver reliable performance, they may impede biodegradability at end-of-life, resulting in notable sustainability concerns. Balancing biodegradability with technical efficacy, longevity, and cost-effectiveness remains a significant obstacle. This chapter reviews the biodegradability of technical textiles, with particular focus on the outcomes of their components in landfill and other disposal scenarios. It highlights current research regarding how raw material selection, production processes, chemical modifications, and environmental factors affect degradation mechanisms. Furthermore, the chapter reviews existing standards for biodegradability testing and explores the compromises between degradability and technical functionality. By synthesizing present knowledge and addressing emerging challenges, the chapter underscores the imperative of integrating environmental responsibility into technical textile design and performance.

Keywords Technical textiles · Biodegradability · Sustainability · Polymer degradation · Biopolymer · End-of-life disposal

Md. S. Ikbal · M. A. Uddin (✉)
Department of Dyes and Chemical Engineering, Bangladesh University of Textiles, Dhaka, Bangladesh
e-mail: abbas.shiyak@dce.butex.edu.bd

Md. S. Ikbal
e-mail: shah.ikbal@dce.butex.edu.bd

T. Kaiser
Department of Wet Process Engineering, Bangladesh University of Textiles, Dhaka, Bangladesh
e-mail: 202113050@wpe.butex.edu.bd

S. S. Muthu (ed.), *Technical Textiles for Sustainability*, Sustainable Textiles: Production, Processing, Manufacturing & Chemistry, https://doi.org/10.1007/978-3-032-29974-1_4

1 Introduction

Technical textiles are engineered materials designed primarily to deliver specific functional performance, such as mechanical, thermal, chemical, or biological, in defined service environments, rather than to fulfil aesthetic objectives. Unlike conventional textiles, which prioritize appearance and comfort, technical textiles are optimized for performance-driven criteria across diverse application sectors, including agriculture, construction, healthcare, transportation, and aerospace [1–3]. The rise of technical textiles has been driven by increasing industrial and societal demand for high-performance materials that meet stringent standards for safety, reliability, and efficiency. Advances in polymer chemistry, nanotechnology, and fiber engineering have enabled the precise control of fiber architecture, morphology, and functional properties to satisfy these performance requirements [4–6].

The global market for technical textiles was valued at approximately $239.01 billion in 2024 and is projected to reach around $253.94 billion in 2025, expanding at a compound annual growth rate between 4.8% and 6.8% annually, driven by sustained demand in the medical, automotive, construction, and defense sectors [7, 8]. This growth underscores the integral role of technical textiles in modern infrastructure and healthcare. For instance, in healthcare, biodegradable polymers enable the fabrication of resorbable sutures, tissue-engineering scaffolds, and implantable devices, which have revolutionized wound care and tissue repair [9–11]. In civil and geotechnical engineering, woven and nonwoven geotextiles facilitate soil stabilization, erosion control, drainage, and reinforcement in roads, railways, embankments, and landfill systems. In transportation, technical fiber composites enhance crash resistance and structural integrity while reducing vehicle weight and fuel consumption [5, 12–14].

The engineering properties that make technical textiles indispensable in modern life also create a fundamental sustainability challenge: functional excellence tends to render these materials environmentally persistent [15, 16]. The vast majority of technical textiles are manufactured from synthetic polymers, which collectively account for nearly 70% of global production [17]. These polymers exhibit negligible biodegradation under natural environmental conditions, with studies reporting degradation rates of less than 5% even after prolonged exposure in soil, freshwater, marine, or landfill settings. Even biopolymers designed for controlled degradation persist in cooler, mesophilic environments where industrial composting conditions are absent [15, 18, 19]. Chemical and physical modifications applied to enhance in-service durability further depress biodegradation potential by reducing microbial accessibility to the polymer chains [20, 21]. Laboratory evidence explicitly demonstrates this trade-off: while untreated cotton can degrade by up to 89% in sewage sludge within 37 days, chemically treated equivalents show a 40% reduction in degradation rates. Synthetic fibers, protected by covalent surface bonds, may show only 2–5% mineralization after extended exposure [15, 18, 21, 22]. Consequently, textile-derived microplastics are now documented at concentrations as high as 5.4 fibers per cubic metre of air and 11.3 fibers per kilogram of soil, demonstrating their pervasive dispersal across environmental compartments and raising serious concerns regarding bioaccumulation and ecotoxicological impacts. These observations underline the need to move beyond durability alone and to explicitly consider how technical textiles behave at the end of life [15, 23, 24].

In this context, biodegradation, defined as the multistage biochemical transformation of complex polymeric materials into simple inorganic compounds (water, carbon dioxide, methane, and biomass) via microbial metabolism in air, water, and soil, has therefore emerged as a strategic design lever for technical textiles [4, 6, 10, 18]. Rather than simply maximizing material persistence, attention is increasingly directed toward engineering polymers that deliver required performance in service yet can re-enter natural biogeochemical cycles at the end of life, thereby mitigating microplastic accumulation and chemical burdens in the environment [5, 24].

The fundamental challenge for the technical textile industry is to align functional excellence with biodegradability, achieving materials that retain required performance, durability, and reliability in use while enabling environmentally compatible degradation after disposal [25–27]. This chapter, therefore, aims to review the current state of biodegradability in technical textiles, particularly, it will explore the mechanisms and influencing factors governing fiber degradation, assess existing biodegradability testing standards, and analyze the trade-offs between environmental responsibility and technical performance. Ultimately, the goal is to highlight the pathway toward harmonizing high performance with environmental responsibility, ensuring that the future of technical textiles is not only functional but also ecologically accountable.

2 Concept of Biodegradation and Biodegradability

The environmental fate of polymeric materials is governed not simply by their durability or origin, but by the mechanisms through which they interact with biological systems after disposal. In this context, biodegradation and biodegradability represent fundamental scientific concepts that describe the biologically mediated transformation of polymers in natural and engineered environments. A clear understanding of these concepts, including their mechanisms and classification criteria, is essential for understanding their environmental fate [28–31].

2.1 Biodegradation and Biodegradability

In the context of polymeric materials, understanding environmental fate requires a precise distinction between degradation and biodegradation as well as between biodegradability and biodegradation. While these terms are often used interchangeably in commercial discourse, they represent distinct physicochemical processes with vastly different ecological endpoints [32, 33].

Degradation denotes an irreversible alteration in a material's macrostructure, characterized by a loss of mechanical integrity, a reduction in molecular weight, or fragmentation. Typically triggered by abiotic stressors such as (UV) radiation (photo-oxidation), thermal stress (thermo-oxidation), mechanical abrasion, or hydrolytic action [34, 35]. These mechanisms cause bond scission and crosslink rearrangement, resulting in material embrittlement, but alone do not equate to environmental elimination [15, 36, 37].

Biodegradation, or biotic degradation, is a specific subset of degradation and a coupled sequence of depolymerization and mineralization driven by microbial metabolic activity. Enzymatic action first cleaves macromolecules into oligomers and monomers, which are subsequently transported into microbial cells and metabolized as carbon and energy sources. For a polymer to be classified as truly biodegradable, it must undergo mineralization, defined as the complete conversion of organic carbon into simple inorganic end products. This distinction highlights a critical principle: while all biodegradable materials degrade, not all degradable materials are biodegradable. A polymer, for instance, that fragments under UV exposure has degraded but not biodegraded, whereas an enzymatically mineralized material may biodegrade without visible fragmentation [15, 38, 39].

Biodegradability is a material property describing the potential for biologically mediated decomposition under defined conditions. It quantifies how completely and how quickly microorganisms can degrade a polymer. The interplay between the polymer characteristics and the receiving environment governs the rate and extent of this process [40–42]. Furthermore, it is essential to distinguish feedstock origin from end-of-life behavior. The term bio-based refers solely to renewable carbon origin, whereas biodegradable denotes susceptibility to microbial mineralization; thus, a bio-based polymer is not inherently biodegradable [43–45]. Biodegradability must therefore always be interpreted relative to specific disposal environments rather than as an absolute material attribute [40, 41, 46].

A common source of ambiguity is the conflation of biodegradability with **compostability**. Compostability represents a regulated, operational subset of biodegradability, defined by international standards [35]. A material is classified as compostable only if it meets specified criteria within a defined timeframe (typically 90–180 days) under controlled thermophilic conditions (50–60 °C), including: (1) $\geq$ 90% mineralization of organic carbon, (2) physical disintegration such that $<$ 10% residue remains on a 2 mm sieve, and (3) absence of ecotoxic effects on plant growth. Thus, all compostable materials are biodegradable under defined composting conditions, but not all biodegradable materials are necessarily compostable [47, 48].

2.2 *Mechanism and Pathways of Biodegradation*

The complete biodegradation of polymer materials is a heterogeneous process involving a cascade of physical and biochemical events. This process is delineated into four sequential stages: biodeterioration, biofragmentation, assimilation, and mineralization, which collectively convert polymeric carbon into inorganic end products and microbial biomass, as illustrated in Fig. 1 [15, 49, 50]. The progression through these stages can be monitored using specific indicators, as summarized in Table 1. These indicators provide quantitative or semi-quantitative measures of the rate and extent of degradation at each stage. Understanding these mechanisms is essential to align performance during the use phase with end-of-life environmental compatibility [51, 52].

Biodeterioration and Abiotic Surface Conditioning
During biodeterioration, the polymer surface is physically and chemically conditioned, making it susceptible to subsequent depolymerization. This stage involves microbial colonization and biofilm formation, supported by extracellular polymeric substances (EPS) that anchor microorganisms to the substrate [55, 57]. Concurrently, abiotic factors generate microcracks, embrittlement, and increased surface roughness, thereby expanding the accessible surface area available for microbial attack [58, 59]. High crystallinity, dense morphology, and hydrophobic or crosslinked finishes suppress biodeterioration and delay downstream biodegradation steps [60, 61].

Biofragmentation
Because high-molecular-weight polymers cannot cross microbial membranes, biofragmentation (depolymerization) is a critical prerequisite for biological assimilation. In this stage, extracellular enzymes such as hydrolases and oxidoreductases secreted by microorganisms cleave polymer chains into oligomers and monomers;

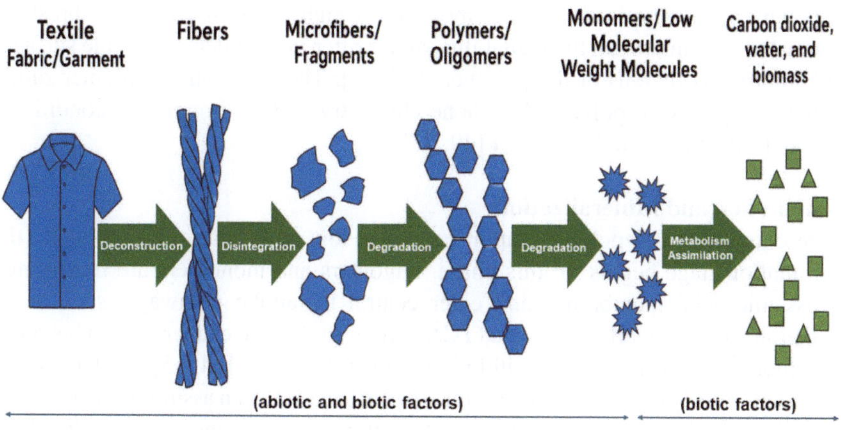

Fig. 1 Schematic pathway of textile biodegradation. Redrawn by the authors from [5]

Table 1 Conceptual stages of polymer biodegradation and corresponding indicators and measurable parameters for quantification of degradation progression

Biodegradation stage	Mechanism	Key indicators	Measurable parameters	References
Biodeterioration	Mechanical weakening and surface damage; microbial colonization, biofilm formation, extracellular enzyme secretion	Surface roughness, biofilm presence, loss of physical integrity	Visual inspection, SEM/AFM, Weight loss, Tensile strength loss	[50, 53]
Biofragmentation	Enzymatic depolymerization of polymer chains	Decrease in molecular weight, tensile strength loss, visible fragmentation	GPC/SEC, tensile strength loss, HPLC/LC–MS analysis	[50, 54]
Assimilation	Uptake of oligomers/ monomers by microbial cells; metabolic incorporation	Substrate utilization, microbial biomass increase, growth rate changes	Biomass quantification, ^{14}C-labeling, protein/lipid (PLFA) analysis	[5, 50]
Mineralization	Complete oxidation (respiration) to CO_2/ H_2O (aerobic) or CH_4/CO_2 (anaerobic)	CO_2 evolution, CH_4 production, organic carbon conversion, and O_2 consumption	Respirometry (CO_2 evolution, BOD/O_2 uptake), ^{14}C/^{13}C isotope labeling, gas chromatography	[55, 56]

abiotic hydrolytic or oxidative reactions may contribute in parallel [10, 50, 62]. Depolymerization typically reduces molecular weight below ~ 10 kDa, a threshold associated with microbial bioavailability, and is often confined to accessible surfaces and amorphous regions of the polymers [18, 49]. The dominant depolymerization pathway depends on polymer backbone chemistry and environmental conditions, and is discussed in detail in Sect. 4 [49, 63].

Assimilation and Mineralization

Once extracellular depolymerization generates low-molecular-mass products, the assimilation stage begins. In this phase, oligomers and monomers are transported across microbial membranes and enter central metabolic pathways, such as β-oxidation and the tricarboxylic acid (TCA) cycle. At this stage, degradation becomes fully biotic, occurring entirely within living microbial cells [64, 65]. Under aerobic conditions, this polymer-derived carbon is partitioned between assimilation, in which a portion is incorporated into new microbial biomass, and mineralization, in which the remainder is respired as CO_2 and H_2O. Under anaerobic conditions, the carbon is transformed into biomass along with a biogas mixture of CH_4 and CO_2 via microbial consortia [55, 66]. The rate and efficiency of these processes depend on the chemistry of the monomer and the metabolic diversity of the surrounding microbiome [67–69].

Mineralization represents the terminal stage of biodegradation, involving the complete conversion of polymeric carbon into inorganic end products, thereby closing the biogeochemical carbon cycle [15, 64, 70]. International standards for biodegradability and compostability measure mineralization through CO_2 evolution or biogas production [71]. While specific regulatory standards may mandate higher thresholds (e.g., 90% for composting), a mineralization level of 60% to 70% in respirometry tests is generally considered sufficient to demonstrate ready biodegradability, with the remaining carbon incorporated into microbial biomass [72–75].

2.3 *Classifications of Biodegradability*

Biodegradability is a context-dependent material property that reflects a polymer's potential to undergo microbial-mediated decomposition under specific environmental conditions. Biodegradability can be systematically classified according to three criteria: (i) the environmental medium, (ii) the redox (metabolic) regime, and (iii) the extent of biodegradation, each of which directly influences the rate and completeness of polymer carbon conversion to inorganic end products (Fig. 2) [15, 39, 63].

2.3.1 Based on Environmental Condition

From a materials science and environmental degradation perspective, the biodegradability of a polymer varies significantly across environmental contexts. These can be broadly categorized by disposal system: terrestrial, aquatic, and industrial/engineered [76, 77].

Terrestrial Environments

The terrestrial environment includes surface exposure dominated by abiotic weathering to full burial within the soil matrix. For subterranean applications, soil constitutes a heterogeneous, biologically complex degradation medium in which aerobic and locally anaerobic microenvironments coexist [10, 15, 61]. Soil burial biodegradability reflects a natural attenuation process driven by microorganisms that rely on intrinsic ecosystem processes to mineralize organic polymers into simpler inorganic compounds. Soil burial typically operates at ambient subsurface temperatures between 18 and 25 °C, where temperature serves as the primary control on enzymatic reaction rates, with microbial activity dropping by approximately half for every 10 °C decrease. In warmer climates or near-surface layers, temperatures

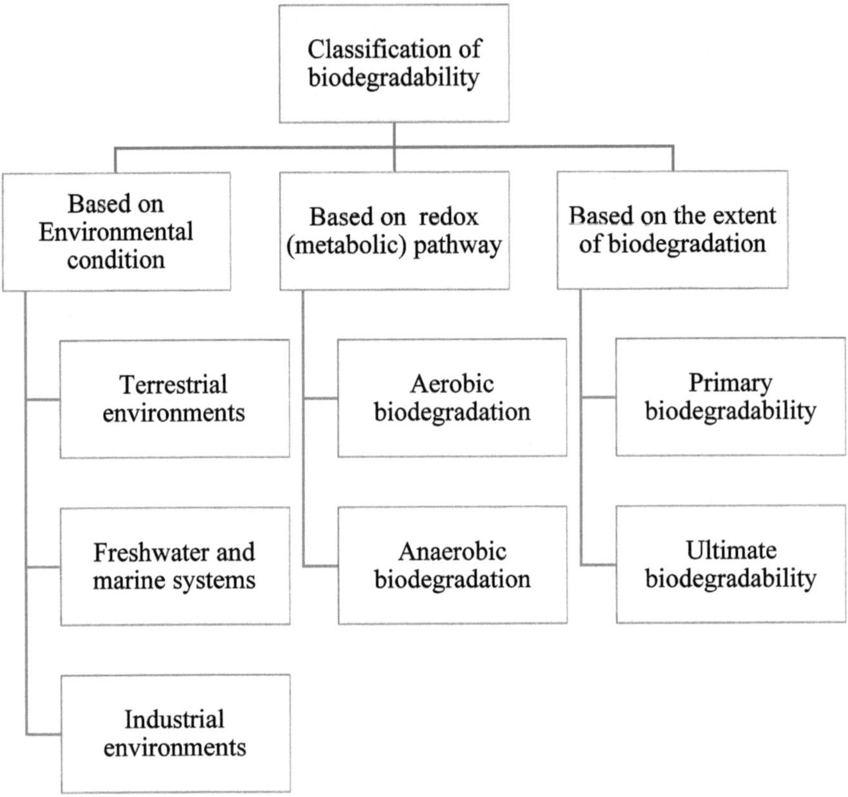

Fig. 2 Classification of biodegradability based on environmental conditions, redox regimes, and extent of degradation

may reach 29 °C, substantially accelerating degradation compared to cooler subsurface conditions. Critical environmental parameters include moisture content, oxygen availability, and pH. Soil moisture should be maintained at 50–70% of water-holding capacity (WHC), which typically corresponds to approximately 20% gravimetric moisture content for active microbial populations. Oxygen availability is highest at the soil-atmosphere interface (0–5 cm depth), decreasing progressively with burial depth. The optimal pH range is 6.0–7.5. Excess moisture restricts oxygen transport and promotes anaerobic conditions, whereas low moisture ($<20\%$) induces microbial dormancy [78–85].

Furthermore, soil heterogeneity, particularly texture and organic matter content, strongly influences biodegradation kinetics by regulating oxygen diffusion and microbial density. Loam and organic-rich soils with high microbial populations support faster CO_2 evolution than compact, clay-rich soils under comparable conditions. The presence of microbial activity, particularly bacteria and fungi, is essential for biodegradability; fungi tend to be the most active agents, often initiating degradation at weak points within the polymer structure, whereas many bacteria require sustained moisture for effective attack [61, 86–89].

Aquatic Environments

Aquatic environments include freshwater and marine systems, which present more restrictive conditions for biodegradability [90–92]. Freshwater ecosystems (rivers and lakes) are low-salinity systems, characterized by diverse microbial communities, fluctuating nutrient levels, seasonal temperatures (4–25 °C), aerobic surface layers, and anaerobic deeper sediments where degradation is driven by hydrolysis and microbial colonization [93–95].

On the other hand, marine environments are high-salinity (3.5% average) systems with high osmotic pressure, alkaline pH (8), oligotrophic conditions, and consistently cold deep-sea temperatures (< 5 °C) that significantly retard metabolic rates [96–98]. Compared with soil or industrial compost (≈ 58 °C), polymer degradation in aquatic systems is strongly constrained by lower, more stable temperatures (< 4 °C in deep waters), dilution of extracellular enzymes and nutrients, and, in seawater, high salinity and ionic strength that inhibit microbial growth and water uptake into hydrophobic polymers [33, 99–102]. Consequently, polymers that mineralize rapidly under thermophilic composting conditions typically degrade by orders of magnitude more slowly in ambient freshwater or seawater [101, 103].

Aquatic biodegradation is further stratified by depth and habitat, creating distinct degradation regimes. In sunlit surface and coastal waters, higher temperatures, UV exposure, and hydrodynamic forcing promote photo-oxidative weathering and biofilm formation, facilitating faster subsequent biodegradation [104–106]. With increasing depth, temperature and oxygen decline, pressure rises, and light is absent; observations from abyssal sites (~ 4150 m) show plastic items remaining essentially unchanged for decades, with no detectable bulk degradation despite microbial colonization. Depth-gradient experiments confirm that even degradable aliphatic polyesters and polyhydroxyalkanoates (PHAs) are decomposed on deep-sea floors at rates that systematically decrease with depth, remaining substantially slower than coastal surface waters [102, 107]. Even within single systems, vertical differences in light, oxygen and resources alter plastisphere communities and functional potential so that polymer fate can differ between surface and deeper layers. Overall, aquatic biodegradation classification highlights markedly slower kinetics and a narrow set of polymer families for which realistic environmental biodegradability is achievable on practical timescales [101, 108].

Industrial/Engineered Environments

Composting and anaerobic digestion (AD) are industrial/engineered biodegradation systems operated with controlled temperature, moisture, aeration, and residence times, in contrast to the stochastic conditions of natural soil or aquatic environments [77, 109–111].

Industrial composting is a controlled, aerobic, bio-oxidative process in which organic matter is converted into stabilized humus through sequential mesophilic and thermophilic microbial activity. It typically operates at thermophilic temperatures of 55–65 °C, with controlled moisture (50–65%), forced aeration, and an optimized carbon-to-nitrogen ratio of approximately 25:1–30:1 [109, 110, 112]. An initial thermophilic phase (40–65 °C) enables hydrolytic cleavage of recalcitrant bonds, followed by mesophilic maturation, resulting in much faster degradation than soil burial or weathering due to intensified microbial activity and optimized conditions [55, 96, 103, 113]. A clear distinction is required between industrial and home composting. Industrial facilities maintain controlled thermophilic temperatures and aeration, whereas home composting generally occurs at lower, fluctuating temperatures and with variable management, leading to slower, less reliable degradation [77].

On the other hand, anaerobic digestion is an oxygen-free, multistage microbial process widely used for treating wet organic waste. AD systems are optimized for pumpable feedstocks and commonly experience mechanical fouling and incomplete degradation when processing intact fibrous materials, making them more suitable for pre-shredded, cellulosic-rich feedstocks than intact, high-performance materials [77, 114–116]. Anaerobic landfill environments represent containment systems designed to minimize environmental leakage rather than maximize decomposition. They develop strictly anaerobic conditions shortly after burial and compaction, producing landfill gas (typically 50–60% methane and 40–50% CO_2) [117, 118]. Landfill internal temperatures are generally mesophilic (25–40 °C) but can reach thermophilic levels (> 50 °C) in deep, biologically active cells; moisture is often limited (typically < 25%) and is a major rate-limiting factor [119–122]. Unlike aerobic composting, such anaerobic, moisture-limited landfill conditions severely restrict the biodegradability of most synthetic polymers, making anaerobic testing important for simulating municipal landfill disposal and evaluating bulky waste materials that are typically landfilled [123].

2.3.2 Based on Redox (Metabolic) Regime

From a sustainability perspective, understanding redox-dependent biodegradation pathways is essential for designing technically sound and environmentally responsible end-of-life strategies for technical textiles. Redox conditions fundamentally influence a polymer's biodegradability by determining which biodegradation pathways are available to microorganisms [15, 33, 68]. The two major redox-dependent biodegradation pathways are illustrated in Fig. 3.

2.3.3 Based on the Extent of Biodegradation

This criterion classifies biodegradability according to the extent to which a polymer is biodegraded, distinguishing primary biodegradability (incipient structural breakdown) from ultimate biodegradability (complete ecological assimilation). This

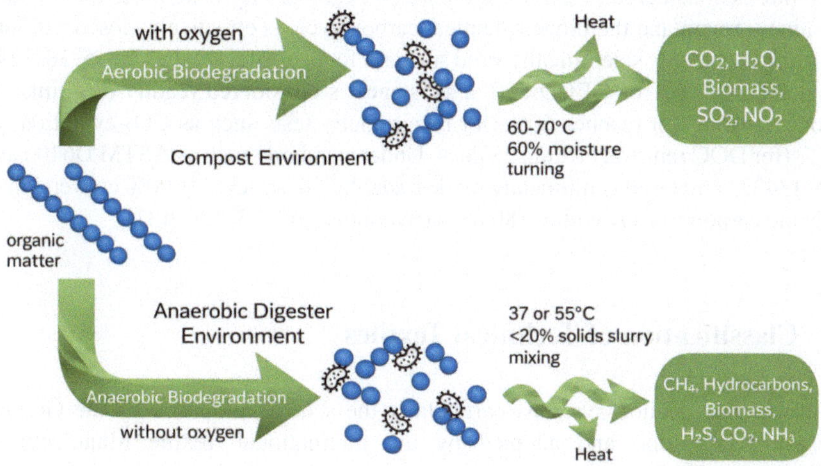

Fig. 3 Comparative biodegradation pathways: Schematic representation of aerobic and anaerobic mineralization of polymeric substrates. Redrawn by the authors from [5]

distinction is critical for avoiding misleading sustainability claims, as many polymeric materials undergo partial structural changes without achieving complete mineralization [124–127].

Primary Biodegradability

Primary biodegradability denotes the initial structural breakdown of a polymer into smaller molecular-weight fragments through microbial activity and/or hydrolytic reactions, leading to the loss of specific physical or functional properties due to molecular weight reduction, without complete elimination of organic carbon from the environment [6, 44, 59]. In laboratory and controlled environmental studies, reported mass loss during primary biodegradation typically ranges from 10 to 50%, depending on polymer chemistry and morphology [33, 128]. Importantly, primary biodegradation does not ensure complete environmental removal. Although the original polymer form may no longer be visible, residual organic fragments, intermediates, or microplastics can persist and, in some cases, exhibit toxicity comparable to or greater than that of the parent material. According to OECD 301 screening guidelines, primary biodegradation is often operationally indicated by a > 20% removal of Dissolved Organic Carbon (DOC) or Theoretical Oxygen Demand (ThOD) [75].

Ultimate Biodegradability

Ultimate biodegradability, also known as ultimate biodegradation or mineralization, is the complete microbial conversion of polymer-derived organic carbon into inorganic end products, including carbon dioxide (or methane under anaerobic conditions), water, mineral salts, and new microbial biomass. In this classification, a polymer is considered to exhibit ultimate biodegradability if no persistent organic residues remain and the biogeochemical carbon cycle is effectively closed and thus constitutes the only scientifically valid standard for eco-safety claims [6, 15, 18, 129]. Under OECD 301 or OECD 310, a substance is considered readily or ultimately biodegradable if it reaches 60% (for respirometry tests such as CO_2 evolution) or 70% (for DOC removal) within 28 days. Under standards such as ASTM D6400 and EN 13432, a material is ultimately biodegradable if it achieves $\geq$ 90% conversion of organic carbon to CO_2 within 180 days (6 months) [48, 75, 130, 131].

3 Classification of Technical Textiles

Modern classification systems, particularly the scheme proposed by the German Textile Association and adopted by the International Textile Manufacturers Federation (ITMF), group technical textiles primarily by twelve application sectors: Agrotech, Buildtech, Clothtech, Geotech, Hometech, Indutech, Meditech, Mobiltech, Oekotech, Packtech, Protech and Sporttech [2, 132, 133].

While alternative schemes based on fiber type, functional properties, or manufacturing technology exist, this application-based framework remains the most widely adopted industrial standard. Accordingly, this application-based classification provides an organizing framework for examining the typical product groups within each sector, the material architectures and chemical systems that constitute them, and the performance requirements that govern material selection and thus determine the feasibility of biodegradation [15, 49].

3.1 Types and Applications

Technical textiles across the twelve application sectors comprise distinct product groups tailored to fulfil specific functional roles in specific contexts. The following overview briefly describes the main technical textile types in each sector and illustrates their typical applications [3, 133].

a. **Agrotech**

Used in agriculture, horticulture, landscaping and forestry for microclimate control, crop protection and soil management. Typical products include mulching and ground-cover fabrics, greenhouse and tunnel cover, windbreak and shading nets, insect-meshes, root-barrier mats, erosion-control blankets, and crop-support twines. These systems operate outdoors, exposed to UV, moisture, and soil contact, often with short functional lifetimes aligned with growing seasons [134–136].

b. **Buildtech**

Applied in building envelopes and civil engineering for sealing, reinforcement and protection. Key products include tensile roofing and façade membranes, flexible formwork liners, reinforcement grids and scrims for concrete and bituminous systems, cavity and roof insulation facings, façade shading fabrics, and temporary scaffolding or weather protection nets. Many Buildtech products are embedded in composite systems and expected to perform for decades [133, 137, 138].

c. **Clothtech**

Comprises functional components integrated into garments rather than fashion fabrics themselves. Major product groups include interlinings, sewing threads, zip and hook-and-loop tapes, waistbands, elastics, waddings, shoulder pads, coated webbings, seam-sealing tapes, and increasingly sensorised or conductive e-textiles in apparel. These components support shape retention, comfort, fastening, insulation and smart functionalities [139–142].

d. **Geotech**

Used in geotechnical and environmental engineering for separation, filtration, drainage, reinforcement and containment. Products include nonwoven and woven geotextiles, geogrids, geonets, drainage composites, geocells, and geomembrane–textile laminates, deployed in roads, railways, embankments, landfills, coastal structures, and erosion-control systems. Geotechnical products are typically buried in soil or aggregates for long periods [13, 143–146].

e. **Hometech**

Encompasses technical components in home and contract interiors. Representative products include upholstery substrates and backing fabrics, mattress ticking and fillings, pillow and quilt waddings, window-treatment linings, carpet backings, and technical filters in appliances such as vacuum cleaners and HVAC units. These textiles must balance comfort and aesthetics with FR, durability and hygiene requirements [3, 133, 147].

f. **Indutech**

Covers textiles used in production processes, machinery and industrial equipment. Important product families include dry and wet filtration media (baghouse and cartridge filters), conveyor belt carcasses and covers, abrasive and polishing backings, expansion joints and bellows, thermal and acoustic insulation felts, process linings, and reinforcement layers in industrial composites. Indutech products often operate at elevated temperatures, in chemical environments, and under high mechanical loads [1, 3, 133].

g. **Meditech**

Includes single-use and reusable textiles in healthcare and personal care. High-volume applications include surgical gowns and drapes, nonwoven masks and caps, wound dressings, bandages, swabs, and absorbent hygiene products (diapers, sanitary pads). High-value products include implantable meshes, vascular grafts, heart valves, tissue engineering scaffolds, and textile-reinforced components in extracorporeal devices. Here, biocompatibility, sterilizability, and controlled interaction with biological tissues are central [133, 148–150].

h. **Mobiltech**

Used in road, rail, air and marine transport systems. Major product groups comprise seat and trim fabrics, headliners, carpets, airbags, seat-belt webbings, tyre-reinforcement textiles, acoustic and thermal insulation layers, trunk liners and lightweight interior composite panels. These textiles face cyclic mechanical loading, temperature fluctuations, UV exposure (where visible) and strict safety and flammability regulations [133, 151, 152].

a. **Oekotech**

Targets environmental protection, remediation and resource conservation. Typical products include air and water-filtration media for industrial and municipal systems, oil-sorbent booms and mats, erosion-control blankets, geocovers for landfills and sludge beds, and barriers for contaminated sites. Their deployment in sensitive ecosystems often motivates the use of renewable and, in some cases, biodegradable fibers [1, 3, 133].

j. **Packtech**

Represents one of the largest sectors by volume. It includes woven and knitted sacks, FIBCs (big bags), leno and mesh bags for produce, tea bag filter materials, bale wraps, container covers, pallet hoods, and composite flexible packaging structures combining textile substrates with films or foils. Packtech products typically have short service lives and highly standardized dimensions [3, 133].

k. **Protech**

Encompasses personal and collective protective equipment. Key applications include firefighter and rescue suits, turnout gear, ballistic- and stab-resistant vests, chemical- and biohazard-barrier suits, arc-flash and molten-metal protective garments, high-visibility clothing, cut-resistant gloves, and protective hoods. Stringent standards for thermal, mechanical and chemical protection define performance [133, 153–156].

l. **Sporttech**

Covers textiles and composites in sportswear, footwear and sporting goods. Products include technical knitwear and woven fabrics for performance apparel, compression and base layers, shoe uppers and linings, tents and sails, as well as fiber-reinforced composites for rackets, bats, skis, boards, helmets and protective padding. Sporttech also includes a growing category of smart textiles that integrate sensors, heating and communication functions into garments [1, 3].

3.2 Material Selection for Technical Textiles

The material selection for technical textiles is inherently application-driven. They typically comprise engineered combinations of fibers, polymer matrices or coatings, and additive systems, rather than single-material constructions. Together, these components determine both the functional performance of technical textiles and their intrinsic susceptibility to degradation processes [15, 49, 157].

3.2.1 Fibers and Textile Substrates

Fibers and textile substrates form the structural backbone of technical textiles and are the primary determinants of their mechanical behavior, processing route and baseline susceptibility to degradation [157, 158]. Four broad fiber families dominate technical textiles:

Natural and Regenerated Fibers
Cotton, flax/linen, jute, coir, hemp, kenaf, wool, and regenerated cellulosic fibers (viscose, lyocell) are widely used where absorbency, conformability, or controlled degradation are advantageous. They form the basis of many erosion-control blankets, biodegradable Agrotextiles, some Oekotech sorbents and filters, interior fillings, and selected medical dressings [157, 158].

Commodity Synthetic Fibers
Poly (ethylene terephthalate) (PET), polyamide (PA 6/66), polypropylene (PP), and acrylics are the workhorses of high-volume sectors such as Geotech, Packtech, Buildtech, Mobiltech, Clothtech, Hometech, and parts of Indutech and Oekotech. Their high strength-to-weight ratio, chemical resistance, thermal stability, and processability in woven, knitted, and nonwoven forms underpin applications ranging from geotextiles and filter media to airbags and interlinings [157, 158].

High-performance Fibers
Aramids (e.g., Kevlar, Nomex), ultra-high-molecular-weight polyethylene (UHMWPE), polybenzimidazole (PBI), glass, carbon and basalt fibers are concentrated in Protech, high-load Geotech/Buildtech reinforcements, Mobiltech safety components and Indutech filtration and insulation, where exceptional tensile, thermal or chemical performance is critical. Their highly aromatic or inorganic backbones are essentially non-biodegradable under environmental conditions [157, 158].

Biodegradable Polymer Fibers
Poly (lactic acid) (PLA), poly (butylene succinate) (PBS), polybutylene adipate terephthalate (PBAT), polyhydroxyalkanoates (PHAs), and their blends with cellulosic or natural fibers are increasingly used in Agrotech, Oekotech, selected Packtech, Meditech devices and experimental Hometech, Sporttech and Indutech products that target industrial compostability or soil biodegradation [15, 49, 159].

3.2.2 Polymer Matrices and Coatings

Many technical textiles function not as standalone fabrics but as components in coated, laminated or composite structures. Typical matrix and coating systems include:

Thermoplastic Matrices
Polyolefins (PE, PP), polyvinyl chloride (PVC), thermoplastic polyurethane (TPU), ethylene–vinyl acetate (EVA), PET, and PA are used as film layers, extrusion coatings, hot-melt adhesives, or bulk matrices in membranes, laminates, and fiber-reinforced composites. Examples include polyvinyl chloride/polyvinylidene fluoride (PVC/PVDF) coated PET architectural membranes, TPU- or PA-coated Clothtech webbings, polyolefin-coated Packtech fabrics, and PP- or PET-matrix Mobiltech and Buildtech composites [160, 161].

Thermoset Matrices
Epoxy, unsaturated polyester, phenolic, and PU resins serve as matrices in structural composites for Protech armor systems, Sporttech equipment, some Buildtech panels, and Indutech components. Their cross-linked networks provide stiffness, dimensional stability and heat resistance, but are essentially non-meltable and non-biodegradable [160, 161].

Bio-based Matrices
PLA, PBS, PHAs, and modified starches are increasingly used as matrices or coating resins in Agrotech, Oekotech, Packtech, and pilot Mobiltech /Sporttech parts, where compostability is a design goal. Natural-fiber/PLA or natural-fiber/PBS composites in mulch mats, erosion blankets, and interior panels exemplify this trend [35, 49, 162–164].

3.2.3 Additives and Functional Finishes

In many technical textile applications, performance requirements are achieved less by altering the base fiber and more by tailoring the chemistry of finishes and additives applied to fibers, fabrics or matrices [165]. Key classes include:

Stabilizers and Protectants
UV absorbers and hindered amine light stabilizers (HALS) in Agrotech, Geotech, Buildtech and Mobiltech protect polyolefins and PET from photo-oxidation and embrittlement [166, 167].

Antioxidants in PP and PE systems slow thermo-oxidative degradation during processing and service [168, 169].

Biocides and anti-mildew agents in Buildtech, Hometech and Protech prevent microbial growth that could cause staining or loss of performance [170–172].

Flame Retardants
Phosphorus-, nitrogen-, and halogen-containing flame retardants are widely used in Hometech upholstery, mattress components, Protech protective clothing, Mobiltech interiors, and certain Packtech applications. They are applied as back coatings, used as intumescent systems, or incorporated into fibers and matrices [154, 165, 173, 174].

Barrier and Repellency Finishes
Durable water-repellent (DWR) and oil-repellent finishes, historically based on fluorocarbons and increasingly on fluorine-free chemistries, are crucial in Clothtech, Protech, Sporttech, and some Oekotech and Packtech products. Hydrophilic finishes or surfactant treatments are used where wetting and filtration are required, such as certain filter media, wipes and hygiene products [171, 175–177].

Antimicrobial and Bioactive Agents
Silver, copper, quaternary ammonium compounds, chitosan, and other antimicrobial finishes are used in Meditech, Clothtech, and Sporttech to control odor or infection risk. In implantable Meditech, bioactive coatings (e.g., heparin, hydroxyapatite, growth factor-bearing polymers) are used to modulate tissue response rather than to enhance environmental durability [88, 165, 170, 173, 178].

Processing aids, Binders and Fillers
Acrylic and PU binders are essential in nonwoven Indutech, Oekotech, and Packtech products; mineral fillers ($CaCO_3$, talc), pigments (TiO_2, carbon black), and antistatic agents are widely used to adjust optical, mechanical, and electrostatic properties across sectors [179–183].

3.3 *Expected Performance Criteria*

Expected performance criteria define what technical textiles must reliably do in service, mechanically, chemically and functionally, before any consideration of end-of-life behavior. These requirements strongly shape fiber, matrix and finish selection, and often favor durability and stability over degradability. The following subsections discuss the key performance expectations across the main application sectors and highlight how they constrain the scope for engineered biodegradation [15, 49].

3.3.1 Mechanical Integrity and Lifetime Performance

Many technical textiles must sustain significant tensile or compressive loads, often over extended service lives under variable stress [1, 3, 6, 184].

High Strength and Low Creep
Geotextiles, Buildtech membranes, Mobiltech safety components and Indutech structural products require sustained mechanical performance over years or decades with

minimal creep deformation. This imperative for long-term dimensional and strength stability inherently favors highly crystalline, chemically resistant polymers that resist both hydrolytic attack and microbial colonization [185–187].

Fatigue and Tear Resistance

Textiles subjected to repeated mechanical cycling (airbags, seat belts, conveyor belts) must withstand millions of stress cycles without failure. The crystallinity and cross-linking required for fatigue resistance simultaneously limit water diffusion and enzymatic accessibility, suppressing biodegradation [5, 6, 15].

3.3.2 Environmental and Chemical Durability

Outdoor and industrial applications expose textiles to environmental stressors that demand robust resistance [3, 6, 15].

UV and Weathering Stability

Long-term exposure to solar radiation, moisture cycling and temperature extremes (Agrotech, Buildtech, Geotech, Mobiltech surfaces) requires durable stabilization systems and coatings. The very durability measures that prevent photo-oxidation and embrittlement during service simultaneously inhibit abiotic hydrolysis and microbial colonization, delaying or preventing biodegradation at end-of-life [15, 188, 189].

Chemical Inertness

Products in contact with fuels, oils, solvents, agrochemicals or aggressive aqueous media (Indutech filters, Protech barrier suits, Geotech liners, some Agrotech and Oekotech systems) must resist chemical degradation. This requirement for chemical inertness directly conflicts with susceptibility to enzymatic hydrolysis and microbial breakdown [39, 190].

3.3.3 Barrier, Safety and Hygiene Function

In sectors where human protection, health, or critical functions are at stake, specific barriers and safety properties dominate the design [6, 165, 184].

Filtration and Barrier Efficiency

Protech garments, surgical drapes and industrial filters must maintain precise porosity and surface properties to exclude pathogens, liquids or particles. The hydrophobic finishes and dense structures that provide these barriers inherently restrict water diffusion into the polymer matrix, a key prerequisite for hydrolytic biodegradation [15, 180, 191, 192].

Flame Retardancy

Hometech upholstery, Protech protective gear and Mobiltech interiors are subject to stringent flammability regulations. The mandated flame-retardant chemistries and inherently stable fibers selected for this purpose can inhibit microbial growth or persist in compost environments, limiting biodegradation [174, 193].

Controlled Biological Interaction
Meditech implants and resorbable scaffolds are designed for predictable, controlled degradation in vivo over defined timescales; permanent implants and most hygiene disposables prioritize stability and safety over environmental degradability, making biodegradation undesirable or impossible [6, 194–196].

4 Biodegradability of Technical Textiles

Building on the conceptual framework established in prior sections, it is now critical to understand how technical textiles actually biodegrade in realistic use and disposal scenarios. The biodegradation of technical textiles is a complex phenomenon governed by the synergy between the material's inherent properties and the surrounding environment. Rather than a single parameter, degradation is driven by multiple interacting factors. To provide a comprehensive analysis, these determinants are categorized into two primary domains: factors based on the textile substrate, which set the thermodynamic and structural limits of degradation, and factors based on biodegradation conditions, which dictate the kinetics and pathways of biological activity [15, 62, 197].

4.1 Factors Based on the Textile Substrate

The biodegradability of any textiles depends on various factors, such as the intrinsic chemistry of the fiber polymer, fiber physicochemical properties, fabrication method, and the chemical treatments applied during processing and finishing. While environmental conditions provide the potential for degradation, the substrate itself determines susceptibility to microbial attack and sets the thermodynamic and kinetic limits of the degradation process. Understanding how these substrate-related factors influence biodegradation is essential for designing technical textiles that maintain performance during use while enabling controlled degradation at end-of-life [15, 49, 51, 147].

4.1.1 Characteristics of Fiber Polymer Materials

The chemical composition of the fiber polymer is the primary determinant of biodegradability, because the backbone structure, functional groups, and chain architecture govern susceptibility to hydrolysis, oxidation, and enzymatic cleavage [15]. However, textile materials exhibit high structural complexity, so degradation behavior can vary widely even among fibers of similar origin. The four fiber families discussed in Sect. 3.2.1, natural/regenerated, conventional synthetics, high-performance materials and emerging biopolymers, also form a useful framework for analyzing biodegradability, as each shows characteristic degradation mechanisms linked to its backbone chemistry and functional groups. Here, the emphasis is not on applications but on how each family behaves in typical end-of-life environments [49].

Natural and Regenerated Fibers

Natural fibers possess inherent biodegradability due to their hydrophilic nature, polar functional groups, and hydrolysable backbone linkages that facilitate microbial adhesion, enzymatic attack, and hydrolytic scission. These "ecologically degradable polymers" encompass natural fibers (cellulosic: cotton, flax/linen, hemp, jute; proteins: wool, silk) and regenerated fibers (viscose/rayon, cellulose acetate, lyocell). Reported biodegradation rates for natural and regenerated fibers are substantially higher than those of petroleum-based synthetics, with cellulosic fibers typically ranging from 10 to 89% under various environmental conditions, whereas conventional synthetics rarely exceed 5% [19, 20, 62, 198, 199].

Cellulose degradation occurs via two parallel mechanisms: oxidation, which cleaves polymer chains and generates new functional groups, followed by enzymatic cleavage of β-1,4-glycosidic bonds by cellulases. The resulting glucose monomers are then assimilated by microorganisms under both aerobic and anaerobic conditions [6, 19, 62, 89]. Untreated cotton textiles exemplify this susceptibility, achieving biodegradation rates of up to 89% in activated sludge systems within 37 days, whereas dyed or finished counterparts degrade more slowly [200, 201].

Protein fibers such as wool and silk degrade more slowly than cellulosic fibers due to their complex interchain cross-linking. Wool is particularly resistant to degradation under dry conditions due to its high sulfur content and disulfide bonds; however, in humid environments, degradation proceeds via fungal and bacterial proteolytic enzymes that target amorphous regions. Silk degradation depends on processing history; raw silk (sericin-rich) degrades faster than degummed silk because sericin serves as a preferred nutrient source for bacteria, demonstrating that surface composition influences microbial colonization [6, 202, 203].

Regenerated fibers exhibit intermediate behavior. Comparative studies show that regenerated cellulose fibers (viscose rayon) degrade more rapidly than native cellulose, owing to their lower crystallinity and higher moisture regain, which increases enzymatic accessibility [49, 201]. By contrast, cellulose acetate degrades markedly more slowly than both cotton and rayon, as acetyl substitution ($-COCH_3$) increases hydrophobicity and suppresses the moisture diffusion required for hydrolytic chain scission [15, 46, 49, 62].

Conventional Synthetic Fibers

Petroleum-based synthetics are recalcitrant to biodegradation due to hydrophobic surfaces, high molecular weight, and chemically stable C–C or aromatic ester linkages that lack recognition sites for common microbial enzymes [5, 49, 204]. Biodegradation in PET and PP rarely exceeds 5%, even under extended exposure in soil, marine, or sludge environments [200]. These polymers typically require abiotic fragmentation, photodegradation or mechanical weathering to reduce molecular weight to oligomeric levels before any enzymatic mineralization can occur [44].

Polyamides (Nylon 6 and 66) exhibit limited susceptibility to specific marine bacteria such as *Bacillus cereus*, *Bacillus sphericus*, *Vibrio furnisii*, and *Brevundimonas vesicularis,* while higher degradation was observed in Nylon 66 than Nylon 6, and both experienced maximal degradation with *B. cereus.* The results concluded that the marine microorganisms have the potential to degrade nylon polymers [6, 205]. In contrast, PET degradation remains limited. Recent developments in biological techniques for the biodegradation of polyester have become a subject of great potential. lipases have induced surface cracks in PET fibers, yet the bulk degradation remains minimal [6, 206].

High-performance Fibers

High-performance fibers, mostly used in technical textiles, such as aramids, carbon fibers, glass fibers, ultra-high-molecular-weight polyethylene (UHMWPE), polyimides, and polyurethanes (PU), are engineered to provide properties that conventional fibers cannot achieve. However, the same precise molecular features that guarantee long service life (rigid backbones, high crystallinity, strong intermolecular bonding) also render these fibers effectively non-biodegradable in natural environments [14, 15, 49, 207].

Polyurethanes exhibit variable degradability depending on the chemistry of their soft segments. Polyester-based PUs can be attacked by fungi (e.g., *Aspergillus niger*, *Cladosporium herbarium*) via hydrolysable ester bonds, whereas polyether-based PUs is far more resistant [38, 208, 209]. Conversely, aramid fibers and carbon fibers exhibit near-complete resistance to biological degradation. Their inertness stems from engineered backbone chemistry: aramids possess rigid aromatic amide linkages, while carbon fibers consist of graphitic carbon sheets. Both structures lack the aliphatic flexibility and accessible functional groups required for enzyme recognition and hydrolysis [49, 210–213]

Similarly, Ultra-High Molecular Weight Polyethylene (UHMWPE) and high-modulus polyethylene (HMPE) fibers resist microbial degradation not due to aromaticity, but due to extreme molecular weight (up to 6 million g/mol) and exceptional crystallinity (> 85%). These physical barriers effectively block the diffusion of water and enzymes into the polymer matrix [214]. Other high-performance fibers, including polyimides (P84), polybenzimidazole (PBI), and polyphenylene sulfide (PPS), are designed for extreme thermal and chemical environments and are therefore not realistic candidates for environmental biodegradation [215–218].

Biopolymers and Blends

To address the persistence of conventional synthetic materials, the industry is increasingly adopting bio-based and biodegradable polymers that combine renewable feedstocks with engineered functionality [5, 43, 147]. These biopolymers are developed with the expectation of providing the eco-friendliness of natural fibers and mimicking the functionality of petroleum-based textiles [49, 219]. Polylactic acid (PLA), the most commercially successful biopolymer, degrades via hydrolysis followed by enzymatic assimilation, though it typically requires industrial composting conditions to achieve timely chain scission [199]. Aliphatic and aliphatic–aromatic copolyesters such as poly (butylene adipate-co-terephthalate) (PBAT) and poly(3-hydroxybutyrate-co-3-hydroxyhexanoate) (PHBH) contain more labile aliphatic ester linkages compared to aromatic esters in PET, enabling faster degradation under composting and elevated-temperature conditions [5, 49, 219, 220, 220]. Similarly, other bioplastics such as polycaprolactone (PCL), polyhydroxyalkanoates (PHA, PHB, PHBV, PHBHx), and polybutylene succinate (PBS) undergo enzymatic degradation in compost, soil, and anaerobic environments [220, 221]. An overview of the characteristics and biodegradation performance of key biopolymers across different environments is presented in Table 2.

Table 2 Characteristics and comparative biodegradation performance of key biopolymers across different environmental conditions (Compiled by the authors from [221, 222]

Biopolymer	Characteristics	Biodegradation performance
PLA	Slightly hydrophobic, UV resistant, high stiffness, thermally stable (≈ 60 °C), composite compound	Compost: 60–70% (30 days) Soil: 0% (120 days) Marine: 3–4% (180 days)
PHB	Brittleness, high level of crystallinity, melting point (10 °C) less than the degradation temperature, low mechanical strength	Compost: 80% (28 days) Soil: 48.5% (280 days) Marine: 38–45% (180 days)
PBS	High flexibility, chemical and thermal resistance, excellent processability, strong composite compound	Compost: 90% (160 days) Soil: 1% (28 days) Marine: 1% (28 days)
Starch Based	Hydrophobic when blended, good mechanical properties, less water permeability, low cost, readily available	Compost: 73.1% (56 days) Soil: 14.2% (110 days) Marine: 100% (50 days)
Cellulose Based	Ductile component, great resistance to scratches, excellent anti-static behavior, high process capability, oil resistance, average mechanical properties	Compost: 100% (18 days) Soil: 100% (180 days) Marine: 75% (150 days)

Blending strategies further modify biodegradability. The degradability of non-biodegradable polymers can often be improved by blending synthetic polymers with biodegradable or natural components. For example, adding starch or other natural fillers to polyolefin matrices increases hydrophilicity and embrittlement and promotes disintegration, although this often represents mechanical fragmentation rather than complete mineralization. A more effective approach combines biodegradable polymers such as PLA, PVA, PCL, or PBS with natural fillers like starch, chitosan, cellulose nanocrystals (CNC), or lignocellulose. These blends enhance water absorption, disrupt polymer continuity, and improve microbial accessibility while reducing persistent microplastic formation [62].

4.1.2 Characteristics of Fiber Properties

Beyond backbone chemistry, the supramolecular structure of textile fibers governs how readily microorganisms and water can access polymer chains [5, 15, 49, 62]. Parameters such as crystallinity, molecular weight, degree of polymerization, intermolecular interactions, moisture diffusion, absorbency, hydrophilicity, surface properties, and level of orientation collectively define the accessibility of polymer chains to hydrolytic and enzymatic attack [49, 62]. In general, polymers that are low in crystallinity, low in molecular weight, and more hydrophilic exhibit faster biodegradation [15, 223].

Crystallinity and Amorphous Regions

Textile polymers are typically semi-crystalline systems comprising ordered, impermeable crystalline domains and disordered, permeable amorphous regions. Biodegradation initiates preferentially in the amorphous zones, where higher free volume and molecular mobility facilitate water diffusion and enzyme penetration [5, 15, 49, 50]. Water diffusion is generally faster in amorphous regions due to their higher free volume than in densely packed crystalline regions. Crystalline regions, conversely, act as physical barriers that retard hydrolysis and oxidation by restricting diffusional transport [50, 224]. Consequently, fibers with lower crystallinity biodegrade more rapidly; for instance, regenerated cellulose (viscose rayon) degrades faster than natural cotton due to its lower crystalline fraction and higher moisture regain [49, 225, 226].

However, when polymers with different chemical structures are compared, crystallinity becomes a secondary determinant. Instead, degradation is primarily governed by chemical functionality, steric hindrance from substituent groups, hydrophilicity, moisture diffusion, and the capacity for swelling and hydrolytic chain scission. Accordingly, hydrophilic and hydrolysable polymers, such as cellulose, degrade more rapidly than hydrophobic, aromatic polyesters, even when the latter exhibit lower crystallinity. Similarly, within a polymer series, biodegradation is controlled by the steric effects of substituents and the degree of substitution rather than structural order [62].

Molecular Weight and Degree of Polymerization

Molecular weight (MW) and degree of polymerization (DP) govern the degradation mechanism. Microbial cells cannot transport high-molecular-weight polymers across their membranes, so degradation must begin with extracellular depolymerization to oligomers and monomers [49, 62, 221]. Higher molecular weight and degree of polymerization, therefore, slow biodegradation by requiring more extensive chain scission and by increasing chain entanglement and packing density. Reducing MW or modifying chain orientation is therefore a primary strategy for enhancing biodegradability in synthetic textiles [43, 49, 62].

Moisture Diffusion and Hydrophilicity

Water acts as the requisite solvent for hydrolysis and the transport medium for enzymes; thus, the hydrophilicity of a fiber, quantified by moisture regain, fundamentally governs the kinetics of chain scission [49, 55]. Hydrophilic fibers swell upon water absorption, exposing internal bonds to enzymatic cleavage, whereas hydrophobic fibers resist wetting, limiting degradation to surface erosion [5, 46].

This hydrophilicity-dependent mechanism explains the significant contrast in observed rates: synthetic fibers with low moisture regain (PET ~ 0.4%, nylon ~ 4%) degrade negligibly compared to hydrophilic cellulosic (cotton ~ 7%, rayon 10–13%) [5, 46, 198, 227]. Even within cellulose derivatives, chemical hydrophobicity dictates performance: cellulose acetate degrades significantly more slowly than rayon because acetyl group substitution blocks the diffusion of moisture essential for hydrolysis. This effect is quantitatively correlated with the degree of substitution (DS): films with DS = 1.7 degraded 80% in five days, whereas triacetate (DS = 3) showed negligible degradation over 28 days [49, 55].

Surface Interactions and Microbial Adhesion

The fiber surface and topography are critical to the initial stages of biodegradation, where microbial adhesion, biofilm formation, and enzyme secretion occur [55, 62]. Surface charge affects electrostatic interactions: most bacterial surfaces are negatively charged, so negatively charged polymers tend to repel bacteria and reduce adhesion, whereas positively charged surfaces favor attachment [228, 229]. Textile preparation processes such as scouring and bleaching increase carboxyl group content, creating more negatively charged surfaces that inhibit microbial colonization [15]. Surface topography further regulates degradation kinetics: rough, porous surfaces with high surface-area-to-volume ratios generally degrade faster than smooth, dense films, as they provide more interfacial area for enzyme adsorption and provide favorable micro-sites for biofilm initiation and nutrient exchange [55, 62].

4.1.3 Method of Fabrication and Textile Construction

Fabrication route and structural configuration influence biodegradation independently of intrinsic polymer chemistry by controlling how environmental agents, such as water, oxygen, and microorganisms, can interact with the materials [6, 15, 49].

Spinning Methods

Conversion of polymer resins or solutions into most textile fibers via extrusion spinning dictates fiber morphology and thereby regulates water diffusion, enzyme adsorption, and degradation rate [42, 230, 231]. Melt Spinning of thermoplastics (e.g., PET, PP, PA, PLA, PHA) produces highly oriented, semi-crystalline, low-porosity filaments with restricted water diffusion and enzyme access [232–235], while wet/gel and dry spinning of polymer solutions (PAN, regenerated cellulose, biopolymers) generate less-oriented, more porous fibers with higher water uptake and internal surface area, favoring faster hydrolytic and enzymatic degradation [42, 236–238], and electrospinning produces sub-micrometer fibers with extremely high surface-area-to-volume ratios, enabling rapid wetting, enzyme adsorption, and markedly accelerated biodegradation relative to conventional filaments [239–241].

Textile Structures

Textile structure further modulates susceptibility. Woven fabrics with looser yarn twist, knitted structures with high porosity, and hydroentangled cellulosic nonwovens all enhance moisture and microbial diffusion, accelerating degradation, while dense polypropylene nonwovens remained highly resistant [24, 62, 198].

Composites

Composite architectures add another layer of complexity to biodegradability. Modern technical textiles often exist as multiphase composites. Their biodegradability depends on matrix type, reinforcement content and interfacial treatment, each of which can independently regulate degradation kinetics [6, 84, 242, 243]. Hybrid composites based on polyethylene or polypropylene matrices reinforced with natural fibers such as jute or coir undergo only partial biodegradation: microorganisms preferentially consume the lignocellulosic fibers, while the synthetic matrices largely remain intact [84, 243]. Biocomposites with biodegradable matrices (PBS, PLA, PCL) reinforced by natural fibers (cotton, jute, flax, sisal) generally degrade faster than neat matrices, as fibers disrupt matrix continuity and increase hydrophilicity. More detailed studies illustrate how interfacial chemistry and fiber content tune this behavior. PBS/cotton composites, for example, degrade more rapidly than neat PBS during the early stages of composting (approximately 0–30 days), but silane-treated cotton fibers slow initial weight loss by improving fiber–matrix adhesion that restricts moisture penetration; yet, both treated and untreated systems eventually reach similar levels of mineralization, indicating that surface treatment modulates the rate but not the extent of biodegradation [242, 243]. Similarly, PBS/jute composites exhibit strong fiber-content and morphology dependence: a composite containing 10% jute lost about 62.5% of its weight after 180 days in compost, whereas higher jute contents reduced degradation due to a denser structure and lower matrix permeability [84, 243]. By contrast, composites incorporating glass, carbon or aramid reinforcements remain essentially non-biodegradable, so these high-performance fibers persist even when surrounding matrices or other components are partially degraded [244].

4.1.4 Surface Characterization of Textiles

Surface characterization and modification of textiles through dyeing, printing, coating, and finishing can influence biodegradability as strongly as the base fiber itself. Surface analyses show that such functionalization often impedes degradation mechanisms by altering surface energy (wettability), topography and chemical reactivity. Many treatments required for technical performance act as physical or chemical barriers to microbial activity, so characterizing the impact of specific surface treatments is critical for predicting the environmental fate of technical textiles [15, 49, 62].

Dyes and Pigments

Dyeing is an essential process in both the conventional and technical textile industries, providing color and aesthetic appeal. However, dyes and pigments can interfere with biodegradation. For example, Reactive dyes, which form strong covalent bonds with cellulose fibers, can reduce biodegradability. The presence of aromatic structures and complex chemical linkages in reactive dyes reduces the polymer's accessibility to microorganisms by blocking enzymatic pathways or hindering water diffusion [20, 21]. Studies show that cotton treated with reactive dyes exhibits up to 32% reduction in biodegradation (seawater, 42 days), while C.I. Reactive Blue 19 reduced degradation by 6% in activated sewage sludge over 102 days [21]. Similarly, studies on viscose rayon show more pronounced effects: undyed viscose degrades at 85% in 60 days, but Remazol Red RB (bifunctional) reduces it to 75%, and Avitera Red SE (poly-functional) further reduces it to 72%, demonstrating that dye functionality inversely correlates with biodegradation [20]. The reduction in biodegradation becomes more pronounced when the dyeing process involves dyes with multiple functional groups that create cross-links within the fiber matrix, further hindering enzyme accessibility and microbial activity [20, 21]. Synthetic pigments and disperse dyes, commonly used in textile dyeing, similarly impede degradation by introducing chemically inert, hydrophobic aromatic structures that block enzymatic sites and resist microbial cleavage [21, 49].

Water-Repellent and Hydrophobic Finishes

Water-repellent finishes such as C6/C8 fluorochemicals, silicones, and waxes block moisture absorption by creating hydrophobic barriers, drastically slowing degradation [5, 21, 85, 245]. C8 perfluoroalkyl sulfonates, historically prevalent but now banned due to bioaccumulation and toxicity, severely inhibited biodegradation by blocking microbial colonization. C6 alternatives remain in use despite persistent environmental concerns [85, 246]. Silicone-based (polysiloxane) and paraffin wax emulsions function by lowering surface energy and forming a physical film around fibers; while less toxic than fluorochemicals, these durable hydrophobic layers significantly retard biodegradation by mechanically preventing water penetration and enzyme adsorption essential for hydrolysis [21, 247]. Cotton treated with water-repellent finishes exhibits only 25% biodegradation after 42 days in marine environments, compared with 72% for untreated cotton, demonstrating a pronounced inhibitory effect of hydrophobic treatments on moisture-dependent degradation pathways [21, 49, 62].

Durable Press and Crosslinking Agents

Durable press finishes often involve crosslinking agents such as imidazolidinone or conventional formaldehyde-based dimethylol dihydroxy ethylene urea (DMDHEU) to improve durability, crease resistance, soil repellency, and dimensional stability. They form covalent cross-links between polymer chains, thereby increasing fiber crystallinity and rigidity while reducing hydrophilicity. Consequently, this structural modification blocks the enzymatic activity, significantly slowing biodegradation [6, 20, 49, 62, 89]. Cotton treated with 5% DMDHEU shows 9% reduction in biodegradation, while imidazolidinone-finished cotton exhibits minor decreases in breaking strength and polymer solubility [21, 89]. Crosslinking creates physical and chemical barriers against enzymatic degradation, reducing water and enzyme penetration and decreasing fiber swelling. In medical textiles (surgical meshes), crosslinking enhances performance (tensile strength > 180 mmHg, infection resistance) but compromises bioresorption kinetics: non-crosslinked biologic meshes degrade in 2–3 months, while crosslinked versions resist enzymatic degradation for extended periods [248–250]. In protective garments, wrinkle-resistant crosslinking finishes reduce tear, tensile strength, and abrasion resistance by 15–30%, 8–38%, and 22–43%, respectively, fundamentally compromising both mechanical durability and end-of-life biodegradability [15, 251].

Flame Retardant Finishes

Flame-retardant (FR) finishes are applied particularly to protective clothing and fire-resistant fabrics. These treatments typically contain halogenated compounds, phosphates, borates, or organophosphates that reduce flammability but simultaneously impede biodegradation [15, 49, 85, 174]. Flame-retardant chemicals alter polymer structure and increase hydrophobicity, thereby reducing microbial accessibility by blocking moisture absorption and enzyme penetration, both of which are critical for degradation in soil and aquatic environments. Consequently, FR-treated fabrics degrade significantly more slowly than untreated fabrics, limiting their suitability for end-of-life environmental breakdown [49, 62, 89].

Antimicrobial and Antifungal Treatments

Antimicrobial finishes are widely used in medical textiles, hygiene products, and sportswear to inhibit bacterial and fungal growth and reduce odors. These treatments often involve incorporating biocidal nanoparticles, such as silver and titanium dioxide nanoparticles, or quaternary ammonium compounds (QAC), which significantly inhibit biodegradation by killing the microorganisms responsible for breakdown and can also increase the hydrophobicity of the fibers [61, 62, 83, 252]. Studies have shown that cotton treated with Ag or TiO_2 nanoparticles exhibits < 10% biodegradation after soil exposure, compared to 72% for untreated cotton; silver nanoparticles alone showed almost no degradation within 25 days, while untreated cotton fabrics exhibited a 50% degradation rate in the same period [20, 49, 252, 253].

Cationic QAC-based agents reduce mineralization to < 10% by forming QAC–fiber complexes that hinder microbial interaction [20, 88]. Antifungal treatments similarly prevent fungal colonization, which is essential for cellulose breakdown in natural fibers and is particularly problematic for outdoor and medical textiles [15, 62].

UV Stabilizers

UV stabilizers, particularly hindered amine light stabilizers (HALS), are essential additives in Agrotextiles, outdoor textiles, construction, and geotextiles to extend service life by absorbing UV radiation and preventing photo-oxidation of polymer chains [254, 255]. These stabilizers protect the material during use by inhibiting UV-induced chain scission and maintaining structural integrity under prolonged solar exposure, while simultaneously preventing the photochemical chain-breaking required for post-disposal degradation [178, 254, 255]. The concentration and chemical structure of UV stabilizers significantly modulate biodegradation kinetics. Studies on PBAT biocomposites demonstrate that properly formulated UV absorbers provide superior aging resistance but exhibit concentration-dependent inhibition of microbial-induced degradation; the type of stabilizer (hindered amine vs. benzophenone vs. benzotriazole derivatives) critically determines whether photo-oxidative or microbial pathways dominate post-consumer degradation [256].

Softening Agents

Unlike most finishes, softening agents enhance biodegradation by increasing hydrophilicity and moisture absorption capacity, thereby facilitating microbial colonization via surfactant-mediated attachment [5, 257]. Cotton treated with amino-functional silicone softeners shows 4% higher CO_2 production in soil over 90 days than untreated cotton, and increased biodegradation from 72% in untreated cotton to 90% with a 5% softener application in freshwater [15, 21]. Reduced fiber stiffness facilitates microbial penetration, in contrast to crosslinking and hydrophobic treatments. The positive impact of softeners on biodegradation is primarily due to reduced fiber stiffness, which facilitates easier penetration of microorganisms into the fiber structure. This contrasts with crosslinking and water-repellent treatments, which decrease biodegradability by forming barriers against microbial action [6, 15, 62].

Surface Pretreatments

Plasma treatments (CO_2, O_2) enhance biodegradation by increasing surface roughness, hydrophilicity, and wettability, making fibers more susceptible to enzymatic attack [62, 258]. CO_2-plasma-treated PLA and PHBV films exhibit significantly faster hydrolysis across pH conditions (except at strongly basic pH), offering pathways to improve biopolymer degradability in medical and textile applications [258]. Similarly, mechanical (particle-size reduction) and thermal modifications can reduce crystallinity, increasing enzymatic accessibility [49, 226].

4.2 Factors Based on Biodegradable Condition

Even substrates with identical chemistry can exhibit very different biodegradation profiles when exposed to different environments. Thus, the intrinsic properties of the substrate (Sect. 4.1) must be interpreted together with the extrinsic conditions that govern real-world biodegradation behavior. In practice, the combined influence of abiotic factors, biotic potential, disposal pathway, and in-service ageing history collectively determines actual depolymerization rates and ultimate mineralization [43, 55, 198].

4.2.1 Environmental Pathways and Disposal Scenarios

As outlined in Sect. 2.3.1, polymer material can enter several distinct disposal environments. For most technical textiles, practical end-of-life scenarios fall into four categories: industrial composting, soil burial, aquatic exposure and anaerobic landfill, each characterized by distinct physicochemical conditions and microbial ecosystems. These differences establish a clear hierarchy of biodegradation potential, ranging from highly accelerated, engineered systems to severely constrained natural settings: industrial composting > home composting > soil burial > freshwater > seawater > anaerobic landfill [43, 49, 62, 84]. In this section, the focus is on how key technical textile materials behave within these environments.

Among these, terrestrial environments represent the most common end-of-life scenario for technical textiles. Within terrestrial systems, degradation occurs primarily through composting and soil burial, both of which rely on microbial activity but differ substantially in environmental control, microbial intensity, and degradation kinetics [15, 259–261]. Composting is a managed, oxygen-rich, and thermally accelerated process, whereas soil burial is uncontrolled, mainly mesophilic, and characterized by heterogeneous aerobic–anaerobic microenvironments [15, 49, 262, 263].

Composting Environment

Composting represents a key end-of-life route for short-lived technical textiles such as disposable medical gowns, hygiene nonwovens, agricultural covers, and biodegradable packaging, where rapid post-use disintegration is essential. It is highly effective for natural lignocellulosic fibers and certified biodegradable aliphatic polyesters, including PLA, PBS, PBAT, and PHAs [129, 221, 264–266]. Under industrial composting conditions, intensified microbial activity promotes ester hydrolysis and chain scission. For instance, PLA-based medical textiles typically achieve more than 90% biodegradation within 180 days, while biodegradable packaging can reach about 95% disintegration within 12 weeks at 70 °C [130, 267].

In comparative studies under similar industrial composting regimes, cellulosic fibers such as cotton attain approximately 89% biodegradation within 37 days, whereas polyester reaches only about 5%, underlining the markedly higher biodegradability of natural fibers relative to conventional synthetics [62, 257]. In contrast, mesophilic home or outdoor composting remains highly effective for cellulosic substrates but markedly less so for synthetic biopolyesters. Cotton and viscose, for example, typically undergo complete disintegration within 60 days under well-aerated, moist conditions, whereas PLA degradation shows strongly temperature-dependent behavior, exceed 60% mass loss within 30 days at 58 °C, yet shows negligible degradation below about 40 °C due to limited ester hydrolysis [268–272].

Composite systems further demonstrate this sensitivity. PBS/cotton-fiber composites achieve 90–95% weight loss within 60–100 days under thermophilic composting (~ 58 °C), whereas suboptimal temperature or insufficient microbial reinoculation can reduce degradation to ~ 65% after three months, highlighting the need for sustained microbial activity and process control [242]. Overall, aliphatic biopolyesters (PLA, PBS, PBAT, PHB/PHBV) and cellulosic substrates achieve high disintegration and CO_2 evolution within 90–180 day cycles under controlled industrial composting [273, 274]. Consequently, composting constitutes a technically and environmentally viable disposal route for short-lifespan, low-load textile applications in Agrotech (such as mulch films, crop covers), Meditech (including disposable gowns, drapes, hygiene products), and Packtech sectors (biobased packaging, filtration media). Conversely, high-performance textiles used in Geotech, Mobiltech, and Protech, based on crystalline aromatic polyesters, polyamides, or polyolefins, remain largely resistant to biodegradation under composting conditions [76, 275].

Soil Burial

Relative to composting, soil burial provides less controlled, predominantly mesophilic conditions, leading to slower and more heterogeneous degradation. Soil burial tests usually expose materials to natural soil (or sometimes compost-amended soil) for defined periods (30–180 days) at ambient temperatures of roughly 20–28 °C, so that biodegradation reflects mixed aerobic and locally anaerobic microniches [10, 162, 276]. Under such conditions, Cotton textiles typically show only 16–27% carbon mineralization to CO_2 after 90 days, while polyester reaches about 13%, indicating limited bioactivity in the tested matrices [19, 55, 62, 257]. Comparative studies show that in more bioactive soils, natural cellulosic materials such as cotton can still lose up to ~ 90% of their weight over four months, yet this remains substantially slower than under industrial composting, and nominally compostable bioplastics like PLA can retain over 99% of their mass after 12 months because thermophilic temperatures required for effective ester hydrolysis are absent [15, 268, 270, 277].

For technical textiles, soil degradation is especially relevant to products designed for ground interaction, such as mulch nonwovens, erosion-control mats, and temporary geotextiles, where both bacteria and fungi contribute to depolymerization and mineralization [10, 36, 86, 268]. In these applications, natural and regenerated cellulosic fibers generally exhibit substantial disintegration and high mineralization over months, whereas polyesters and polyolefins show only limited surface oxidation or bioerosion with minimal bulk carbon conversion [19, 278, 279]. Soil burial, therefore, remains a critical environmental pathway for evaluating terrestrial biodegradability of ground-contact technical textiles [10, 36].

Aquatic Environments: Freshwater, Seawater, and Wastewater Treatment

Aquatic environments are critical pathways for technical textiles used in marine ropes, fishing gear, coastal geotextiles, filtration media, oil spill containment and cleanup, buoyancy devices like life jackets and rafts, and performance apparel, which are prone to microfiber shedding and represent a significant source of global microplastic pollution [90–92]. Compared with compost and soil, freshwater and seawater generally exhibit lower temperatures and microbial biomass; seawater additionally imposes elevated salinity. As a result, many polymers that are readily compostable show only limited breakdown in aquatic systems [15, 55, 62, 90].

Freshwater Environments

In freshwater, cellulosic fibers such as cotton and rayon generally show high biodegradability, often exceeding 70% mineralization in lake water within several weeks and performing similarly in activated sludge [90]. Polyester and other conventional synthetic polymers, by contrast, exhibit negligible biodegradation in these tests despite prolonged exposure. For instance, studies reported 77% biodegradation of cotton and 73% of rayon in freshwater after 35 days, whereas polyester showed almost no measurable degradation [20, 21, 90].

Natural fibers such as cotton and jute can lose approximately 50–70% of their mass within about 90 days in freshwater, whereas PLA and other bioplastics remain largely inert because the thermophilic temperatures required to drive ester hydrolysis are absent. Empirical studies further show that cellulosic fibers lose mechanical integrity over weeks in surface waters, whereas conventional technical fibers such as PET, PP and PE, and highly crystalline PLA, often retain their structure for many months to years, experiencing only slow surface weathering and mechanical abrasion with very limited overall mass loss [15, 21, 261, 280]. This contrast confirms that industrial compostability (e.g., PLA or PBAT) does not equate to aquatic biodegradability. By exception, certain aliphatic biopolyesters, particularly PHAs and PHBV copolymers, can reach high, sometimes near-complete mineralization in marine or aquatic assays, indicating that specialized microbial consortia can effectively colonize and degrade these chemistries [102, 163, 281, 282].

Marine Environment

In marine environments, degradation is generally slower than in freshwater environments, reflecting the combined effects of salinity, lower nutrient availability, and often reduced oxygen, all of which constrain microbial growth and enzymatic activity. In one study, cotton and rayon showed only about 50% degradation in seawater, where they exceed 70% in lake water, while polyester remained essentially non-biodegradable in both media [21, 90, 283]. Cotton fibers can lose 60–80% of their tensile strength over six months in seawater, yet complete mineralization often requires more than a year, and only a narrow class of polymers, most notably PHAs, show appreciable mass loss (around 30–50%) within similar timescales [21, 163, 277, 284].

Wastewater Treatment Plant (Activated Sludge)
Wastewater treatment plants represent the most aggressive aquatic pathway for biodegradation and are widely used to simulate the fate of textile microfibers released during domestic laundering as they transit municipal treatment plants. These engineered systems have microbial densities that are orders of magnitude higher than those in natural waters and tightly controlled dissolved oxygen, pH, and mesophilic temperatures (20–35 °C) [21, 75, 163]. In this matrix, cellulosic fibers such as cotton, viscose, and rayon typically achieve 80–95% mineralization within about 28 days, driven by abundant cellulase-producing bacteria, whereas conventional synthetic polymers such as polyester and nylon show less than 1% biodegradation and are predominantly removed via physical capture in sludge rather than chemical breakdown [15, 21, 285].

Biodegradation extents in activated sludge are consistently greater than in other aquatic media. At relatively low total suspended solids (TSS) (~ 30 ppm), activated sludge supports greater overall biodegradation of cellulosic materials than lake water or seawater, with cotton and rayon yarns (without finishes) achieving more than 70% biodegradation in both activated sludge and lake water, but only approximately 50% in seawater [15, 90, 257, 283].

Anaerobic Environments: Landfill Conditions
In anaerobic landfill conditions, conventional technical textiles such as polyester and nylon, as well as nominally compostable PLA, can remain virtually unchanged for decades, with landfill studies indicating near 100% mass retention even after 30 or more years due to the combination of oxygen absence and insufficient thermal triggers for their degradation pathways [270, 277, 286]. Since roughly 80–85% of global textile waste is ultimately disposed of in landfills, landfill-focused investigations are essential for assessing both the biogenic methane potential (BMP) of natural fibers and the long-term persistence of synthetic polymers [260, 261, 287].

Even under optimal anaerobic conditions, natural fibers decay far more slowly than under aerobic composting. Recent data suggest that cotton may require approximately 10–20 years to reach approximately 50% degradation in a dry tomb landfill, compared with months under industrial composting or active soil/compost conditions. For technical textiles, this means that landfill disposal effectively locks in the persistence of most synthetic and many biobased materials, regardless of their performance in aerobic tests. [15, 68, 257, 288].

4.2.2 Abiotic Environmental Factors

Abiotic factors are non-living physical and chemical elements that initiate structural weakening (biodeterioration) and precondition them for subsequent biotic attack. They operate independently of microorganisms, yet critically condition textile degradation by modifying polymer structures and the surrounding medium. The dominant abiotic drivers across terrestrial, aquatic and engineered systems are temperature, solar radiation, water availability and salinity, pH and nutrient balance, and oxygen transport/porosity [49, 50, 55, 62].

Temperature

Temperature is one of the primary factors impacting both the rate of chemical (abiotic) and microbial (biotic) degradation [6, 19, 55, 61, 289]. In terrestrial environments, mesophilic temperatures (20–30 °C) generally support active biodegradation in soils, whereas industrial composting operates at elevated thermophilic temperatures (45–60 °C, typically near 58 °C) to sustain thermophilic consortia and accelerate polymer breakdown [15, 49]. Laboratory protocols frequently select fixed "high-normal" temperatures (about 15–30 °C in aquatic tests and 20–28 °C in terrestrial tests) to standardize conditions and shorten test durations, even though these may overestimate biodegradation relative to cooler natural settings [6, 15].

Rising temperature generally increases reaction rates and microbial growth, provided moisture and nutrients are adequate [20, 55, 61, 252]. Warmer environments thus favor higher microbial biomass and faster biodegradation, while low temperatures suppress growth and enzyme activity, significantly slowing decay [15, 198]. This pattern is evident in cold marine systems, where synthetic fibers such as polyester degrade very slowly, whereas natural fibers (e.g. cotton) show faster breakdown in warmer waters due to more intense microbial colonization [62]. Quantitatively, one seawater study reported a 42% decrease in biodegradation when the test temperature was reduced from 20 to 7 °C, illustrating the sensitivity of microbial efficiency to temperature [5]. Thermal energy also acts at the polymer scale by increasing molecular mobility. When the temperature approaches or exceeds the glass transition temperature Tg, segmental motion and free volume increase, which enhances water diffusion into the matrix and facilitates both chemical scission and enzymatic attack [15, 49, 55].

In controlled composting, high temperatures are deliberately used to intensify degradation. For example, PBS composites tested under ISO 148552 at 58 °C showed substantial mineralization [242], and jute-reinforced polymer composites exposed to a stepped regime (initially set at 35 ± 5 °C, raised to 58 ± 5 °C for 4 days, then kept at 50 ± 5 °C for 28 days before returning to 35 ± 5 °C) exhibited efficient disintegration under the resulting thermophilic conditions [84]. Such profiles promote rapid activity of thermophiles, although the thermophilic phase can temporarily reduce overall microbial diversity [290].

Soil burial tests have shown that biodegradation rates are higher when experiments are conducted in summer (average temperature ≈ 25 °C), reflecting the temperature dependence of soil microbial processes [252]. Aquatic biodegradation experiments often use a constant temperature, typically around 25 °C, to standardize conditions for aerobic biodegradation [21, 90].

UV Radiation

Exposure to ultraviolet (UV) radiation from sunlight induces photo-oxidation of textile fibers, generating free radicals, chain scission and surface cracking. This embrittlement increases surface area and porosity, often facilitating microbial colonization and enzymatic degradation during subsequent disposal. For natural polymers such as cotton and linen, UV-induced changes in hydrophilicity and surface roughness can enhance water uptake and make fibers more susceptible to cellulolytic enzymes, whereas in synthetic polymers like polyester, UV primarily acts as a pre-conditioning step that may accelerate or simply fragment the material depending on stabilizer content [15, 49, 62].

Hydrolysis, Moisture Content, and Salinity

Water availability is a prerequisite for both aerobic and anaerobic degradation because it controls the cleavage of susceptible bonds and the transport of nutrients and enzymes in the surrounding medium [55, 198, 289]. In hydrolysable polymers such as aliphatic polyesters, diffusion of water into the bulk triggers ester cleavage and molar-mass reduction, generating oligomers that microorganisms can assimilate [43, 49, 55] The rate of this preliminary hydrolysis is strongly coupled to how effectively the textile matrix can interact with the aqueous phase [21, 90]. The role of fiber-level hydrophilicity and moisture regain is discussed in Sect. 4.1.2; here, the focus is on system-scale water availability.

In composting, intermediate moisture levels are required to balance oxygen supply and water activity: values around 45–65% are generally suitable for microbial activity, with a narrower optimum near 55–65% often reported for industrial composting [55, 290]. Above this range, pore spaces become waterlogged, and local anaerobic zones develop, slowing aerobic biodegradation, whereas below about 40% moisture, desiccation limits microbial growth and inhibits hydrolytic enzymes [49, 84]. Soil burial tests, therefore, maintain controlled humidity, typically around 60 $\pm$ 5% of maximum water holding capacity or via periodic irrigation, to prevent moisture-limited microbial activity and hydrolysis [84, 88, 89, 253].

In aquatic environments, bulk water is abundant, but salinity becomes a critical modifier of both water–polymer interactions and microbial performance. Elevated ionic strength in seawater can reduce water ingress into polymer matrices and impose osmotic stress on many degraders, leading to systematically lower biodegradation rates in brackish and marine waters than in freshwater [10, 223]. Consequently, natural and regenerated cellulosic fibers generally degrade faster in low-salinity or nutrient-enriched conditions, whereas synthetic fibers show very limited breakdown in saline systems [62, 283]. The behavior of poly-L-lactic acid (PLLA) further illustrates this salinity effect: PLLA has been shown to degrade faster in distilled water than in seawater, because the absence of dissolved mineral salts facilitates water penetration and accelerates ester hydrolysis [23].

pH, C/N ratio, and Nutrient Availability

The chemical milieu, particularly the pH level, carbon to nitrogen (C/N) balance, and macro- and micronutrients, has a significant influence on microbial activity that can either accelerate or suppress polymer breakdown [6, 43, 49, 198].

Biodegradation is highly sensitive to pH because microbial community structure and enzyme activity both exhibit relatively narrow optima. Bacteria usually proliferate best at near-neutral pH (6–8), whereas many fungi tolerate wider ranges [6, 55]. In composting, pH often drops during early acidogenesis due to organic acid formation and then rises toward neutrality or mild alkalinity as ammonification progresses; overall degradation tends to be maximized between pH 5.5 and 8.0 [21, 90, 290]. Acidic conditions below pH 6 can reduce cellulase activity and slow the breakdown of natural fibers, whereas more alkaline conditions (pH > 8) may favor surface erosion and hydrolytic cleavage of some polyester-based fibers [49, 62]. In aquatic systems, modestly lower pH can enhance hydrolytic degradation of susceptible polyesters; for instance, PLLA has been shown to degrade faster in slightly acidic distilled water than in more alkaline seawater [23]. Soil burial tests typically range from pH 4 to 7.5 when commercial composts are used [88, 89, 253].

The C/N ratio is essential for biodegradation, especially during composting and in soil. The ideal range is 15:1–30:1 because microbes need carbon for energy and nitrogen for biomass synthesis and growth [55, 291]. Too high a C/N ratio restricts nitrogen availability, decreasing microbial activity and enzyme production, which slows degradation. Conversely, too low a C/N ratio results in excess nitrogen that can be lost as ammonia, potentially causing process instability or odors, thus reducing efficiency [290]. Keeping the C/N ratio at an intermediate level ensures a balanced supply of carbon and nitrogen, promoting continuous microbial growth and steady breakdown of natural fibers and biodegradable polymers [15, 49, 62].

Beyond carbon and nitrogen, microorganisms require adequate supplies of both macro (e.g., P, K) and micronutrients (e.g., Fe, Mn, Cu, Zn, B) to sustain metabolic activity and enzyme systems relevant to polymer breakdown [49, 198, 290]. In nutrient-rich soils and composting environments, natural fibers such as cotton and wool consistently exhibit faster biodegradation, reflecting increased microbial biomass and metabolic activity [15, 49, 62]. Consequently, standard soil burial protocols specify substrates containing readily available nitrogen, phosphorus, potassium, and trace metals to minimize nutrient constraints during biodegradation testing [21, 252]. Similarly, aquatic biodegradation tests commonly incorporate defined nutrient solutions, such as ammonium chloride (NH_4Cl) and potassium dihydrogen phosphate (KH_2PO_4), to prevent nutrient limitation and ensure microbes have sufficient resources for efficient biodegradation [90, 283].

Oxygen Flow and Porosity

Oxygen availability determines whether aerobic or anaerobic pathways predominate, thereby shaping microbial assemblages, degradation rates, and end products. Well-aerated systems such as compost, surface soils, and oxygenated waters support aerobic metabolism, which converts organic carbon to CO_2, H_2O, and biomass; effective composting typically maintains pore-gas oxygen levels above ~ 10% [6, 10, 49, 290]. In aquatic tests, oxygen uptake or CO_2 evolution is therefore used to quantify ultimate aerobic biodegradability [21, 90]. Where oxygen is depleted, as in deep sediments or landfills, anaerobic consortia dominate, and organic matter is converted primarily to CO_2 and CH_4, typically at much slower rates than in aerobic systems [5, 6, 19].

Material porosity and particle-size distribution influence how well oxygen can reach compost piles or soil matrices. Smaller particles offer more surface area for microbes to grow, but if they become too fine, they can restrict airflow and water movement, leading to localized anaerobic zones that shift degradation towards the methanogenic pathways [55, 290].

4.2.3 Biotic Factors on Biodegradation

Biotic factors involve the living organisms and their biological processes that ultimately drive final mineralization once abiotic processes have rendered polymers accessible to biological attack. While microorganisms are the primary biotic agents, other critical factors, including extracellular enzymes, microbial community dynamics, and biofilm formation, collectively govern the actual biodegradation of textile materials [5, 37, 49, 292–294].

Microorganisms, including bacteria, fungi, and algae, are the essential drivers of biodegradation. These organisms secrete extracellular enzymes that cleave high-molecular-weight polymers into soluble oligomers and monomers [5, 49, 201, 221]. Fungi are often more effective than bacteria in degrading lignocellulosic fibers such as cotton, whereas bacterial activity generally requires higher moisture levels to achieve comparable breakdown [89]. Consequently, the type and origin of the inoculum, whether activated sludge, compost, or native soil microflora, strongly govern enzymatic capability, population density, and observed degradation rates [6, 15, 200]. Standardized biodegradation tests frequently employ highly active, non-native inocula under optimized conditions, which often leads to overestimation of environmental biodegradation. In contrast, studies that rely on naturally occurring microbial communities in less enriched media tend to show slower and less complete degradation [15].

Biodegradation involves specific types of enzymes. Hydrolases, such as lipases, esterases, and cutinases, cleave ester bonds in plastics such as PLA, PBS, and PET. Oxidoreductases, including laccases and peroxidases, attack more resistant structures, such as polyethylene chains and aromatic compounds, via oxidation. In cellulosic textiles, cellulases hydrolyze glycosidic bonds into assimilable sugars. Therefore, whether a particular textile material can biodegrade effectively depends largely on the presence of these enzyme systems in the environment [292, 295–297].

4.2.4 Influence of Service Life on Biodegradation

Service-life conditions act as a preconditioning stage that can either enhance or impede the end-of-life biodegradation of technical textiles by modifying polymer chemistry, morphology, and fragmentation state prior to disposal [15, 49]. During use, exposure to ultraviolet radiation, mechanical abrasion, temperature fluctuations, humidity, and chemicals can induce photo-oxidation, chain scission, embrittlement, surface cracking, and fiber fragmentation. These changes increase the surface area, porosity, and water uptake, making polymers, especially cellulosic substrates and some aliphatic polyesters, more accessible to hydrolysis and microbial colonization once they enter compost, soil, or aquatic environments [49, 62]. Textiles used in high-stress applications such as sportswear, agrotextiles, and industrial fabrics often reach end-of-life in a partially damaged state due to prolonged exposure during service. This pre-damage can accelerate subsequent biodegradation when material chemistry and disposal conditions are favorable. In contrast, textiles from low-stress applications tend to retain more intact and smoother fiber. When these textiles are eventually discarded, their less-damaged polymer chains make them more resistant to degradation, even when they experience the same abiotic and biotic conditions as more heavily used materials [15, 49, 62].

5 Assessment of Biodegradability

The intrinsic and extrinsic factors governing the biodegradation of technical textiles, as well as the quantification of these processes, require standardized assessment frameworks. Evaluating the biodegradability of technical textiles is essential for validating sustainability claims, ensuring regulatory compliance, and understanding material fate in post-consumer environments. Consequently, the fundamental concepts of biodegradation testing methodologies across diverse environments are examined, alongside the specific challenges that arise when applying existing standards to engineered, high-performance textiles [15, 41, 298].

5.1 Testing of Biodegradability of Technical Textiles

Evaluating the biodegradability of technical textiles is essential for validating their sustainability claims and positioning them within circular economy frameworks, yet this is intrinsically complex because these are engineered materials specifically designed for functional performance, which frequently conflict with rapid biodegradation [6, 147, 299]. Biodegradability cannot be inferred from surface damage, color change, or loss of mechanical strength alone; it requires demonstrating that the material progresses through sequential stages of biodeterioration, biofragmentation, assimilation, and, ultimately, mineralization as outlined in Sect. 2.2 [50, 53, 56].

In practice, this multistage process is monitored using a combination of analytical endpoints that track both the extent of mineralization and the related physical and chemical changes [6]. Current biodegradability assessment frameworks for textiles rely predominantly on standards originally developed for plastics, established by the International Organization for Standardization (ISO), the European Committee for Standardization (CEN), the American Society for Testing and Materials International (ASTM), the Australian Standards (AS), and other standardizing bodies. These standards were established to meet regulatory requirements and have been adapted for textiles; however, they did not explicitly consider the structural complexity and functional requirements of textile materials [43, 257].

Within this standardization framework, test methods are generally divided into two broad categories. The first comprises direct "ultimate biodegradation" assays, such as aerobic and anaerobic respirometry tests, and specialized ^{14}C-labelling protocols that quantify CO_2 evolution, O_2 consumption, or biogas production and provide direct evidence of mineralization by tracking the conversion of polymer carbon into metabolic end-products [72, 73, 289, 300, 301]. The second category includes indirect techniques, like weight loss measurements, tensile strength retention, molecular weight reduction determined by gel permeation chromatography or size exclusion chromatography (GPC/SEC), as well as spectroscopic and thermal analyses using Fourier-transform infrared spectroscopy (FTIR), thermogravimetric analysis/differential scanning calorimetry (TGA/DSC), and X-ray diffraction (XRD) alongside scanning electron microscopy (SEM) to examine surface erosion and biofilm growth. These methods mainly assess physical and chemical disintegration, which may precede or occur without complete mineralization [21, 83, 242, 302]. Consequently, claims of biodegradability must be interpreted with careful consideration of both the analytical endpoints applied and the standards under which testing is conducted, as indicators such as weight loss, surface cracking, or mechanical deterioration alone cannot be taken as evidence of actual biodegradability or mineralization [303, 304].

5.2 *Testing Methodology and Environments*

While Sect. 4.2 describes how environmental conditions influence biodegradation behavior, standardized testing methodologies are required to quantify performance in a reproducible and comparable manner. These methodologies focus on measurement endpoints, test validity criteria, and compliance with international standards, rather than on environmental mechanisms [15, 41, 286].

Methodological Classification of Biodegradation Tests
Biodegradation tests are implemented in three principal methodological formats, each designed to simulate environmentally relevant disposal scenarios or controlled conditions that replicate fundamental degradation mechanisms. These approaches differ in their capacity to reproduce natural complexity while maintaining standardized conditions for quantitative mineralization measurement: laboratory tests using defined media, which provide controlled and reproducible systems for quantifying biochemical degradation; simulation tests reproduce key features of disposal environments while maintaining experimental control; and field tests expose materials to natural environments (in situ) but offer limited analytical precision [6, 41, 50, 298].

Among these, laboratory and simulation tests form the basis of most regulatory frameworks and certification schemes, as they allow direct comparison with reference materials and theoretical conversion limits. Across these methodological approaches, four environments predominate in textile biodegradation testing: industrial and home composting, soil, aquatic systems, and anaerobic (landfill-like) conditions. Additionally, weathering is used as a complementary preconditioning step to evaluate how abiotic aging influences subsequent biological degradation. These scenarios yield divergent degradation kinetics and mechanisms and therefore require scenario-specific assessment, typically using the methods above to represent distinct disposal routes [84, 243, 289, 305]. The specific test methods/standards, test environments, biodegradation indicators and validity/pass criteria corresponding to these methodological and environmental categories are summarized in Table 3.

Laboratory Tests Using Defined Media

Laboratory-based biodegradation tests are conducted in controlled systems using synthetic or defined media inoculated with standardized microbial consortia. These tests quantify ultimate biodegradation by measuring oxygen consumption, carbon dioxide evolution, or biogas production relative to theoretical values such as theoretical oxygen demand (ThOD) or theoretical carbon dioxide ($ThCO_2$). Respirometric techniques are central to these methods and enable continuous monitoring of mineralization under aerobic or anaerobic conditions, with test validity established by verifying adequate degradation of a reference material. While highly reproducible and essential for material screening and regulatory approval, these tests are not intended to represent real disposal environments directly [41, 334].

Simulation Tests Representing Disposal Scenarios

Simulation tests are designed to evaluate biodegradation under predefined disposal routes, using real or synthetic media while maintaining controlled operational parameters. They constitute the core methodologies referenced in international biodegradability standards and form the basis for certification schemes, because they combine environmental relevance with quantifiable mineralization or disintegration endpoints [6, 41, 56, 334]. In practice, simulation testing for technical textiles focuses on four main scenarios: industrial and home composting, soil exposure, aquatic systems, and anaerobic conditions such as digestion or landfill.

Aerobic Composting Test Methodologies

Industrial and home composting tests represent the most established pathway for certifying compostable textiles. They primarily quantify biodegradation under thermophilic, well-aerated conditions using CO_2 evolution as the primary indicator and underpin compostability certification schemes for disposable textiles. Standards such as ISO 14855-1/2 and ASTM D5338 define laboratory-scale industrial composting tests under controlled aeration and thermophilic conditions; test validity is established by confirming sufficient biodegradation of a reference material (typically cellulose), while pass criteria for compostability claims are specified in product

Table 3 Overview of standardized biodegradation test methods for assessing biodegradation and disintegration of textile materials in composting, soil, aquatic and anaerobic environments

Test method/ standard	Environment + conditions	Test type	Biodegradable indicators	Test validity and criteria	References
ISO 14855-1/ 2/ASTM D5338	Industrial compost, 58 °C, controlled aeration	Laboratory	CO_2 evolution (respirometry or gravimetric)	**Test Validity**: Reference material (cellulose) must degrade > 70% after 45 days **Pass Criteria (per EN 13432 & ASTM D6400 Specification)**: > 90% of organic carbon converted to CO_2 within 6 months	[48, 74, 130, 306, 307]
ISO 20200	Lab-scale compost, 58 °C	Laboratory	Physical disintegration	**Pass Criteria:** > 90% of material passes 2 mm sieve within 45–90 days (screening test for disintegration; does not measure mineralization)	[308]
ISO 16929	Pilot-scale compost, 45–75 °C	Simulation	Physical disintegration	**Pass Criteria:** > 90% of material passes a 2 mm sieve after 12 weeks in a realistic environment (screening test for disintegration)	[309]
ISO 21701	Accelerated hydrolysis + compost, 58 °C	Simulation	Hydrolysis + CO_2 evolution	**Pass Criteria**: CO_2 evolution profile must be comparable to reference material; claims of compostability require compliance with EN 13432 or equivalent specification	[48, 310]
ISO/CD 17952	Simulated compost, 58 °C	Laboratory	Mechanical/ surface area loss	**Pass Criteria**: Under development. Intended to quantify strength loss and surface area reduction for textiles	[311]
AS 5810/ NF T51-800	Home compost, ambient mesophilic conditions (typically ≈ 20–30 °C)	Simulation	CO_2 evolution, disintegration	**Pass Criteria:** > 90% biodegradation (12 months) and > 90% disintegration (180 days) at ambient temperatures	[312, 313]
ISO 17556	Controlled soil, 20–28 °C, 40–60% WHC	Laboratory	O_2 demand / CO_2 evolution	**Test Validity:** Reference material must reach > 60% biodegradation (6 months) **Pass Criteria (per EN 17033)**: > 90% conversion to CO_2 within 24 months	[314, 315]

(continued)

Table 3 (continued)

Test method/ standard	Environment + conditions	Test type	Biodegradable indicators	Test validity and criteria	References
ASTM D5988	Controlled soil, 20−28 ± 2 °C, 80–100% WHC	Laboratory	CO_2 evolution	**Test Validity**: Reference material must evolve > 70% of theoretical CO_2 within 180 days **Pass Criteria**: Equivalent to ISO 17556	[316]
ISO 117211/2	Soil burial, 29 ± 1 °C, 95–100% RH (Part 2 uses extended duration)	Laboratory	Part 1: Tensile strength loss Part 2: Tensile strength retention	**Test Validity**: Control cellulose fabric must lose ≈ 80% tensile strength (9-day maximum) **Pass Criteria**: Part 1: percentage of strength loss against a control (Comparative method only for rot susceptibility, no single pass/fail threshold); Part 2: Rot-retardant textiles must retain ≥ 75% of initial tensile strength after extended burial	[317, 318]
AATCC 30 Test I	Soil burial, 25 ± 5% moisture	Laboratory	Breaking strength retention, fungal growth	**Test Validity:** Breaking strength must be measurable both before and after burial (2–16 weeks in soil) **Pass Criteria**: Comparative method only—no single pass/fail threshold, higher strength retention indicates greater resistance	[319]
ISO 14851/ ISO 14852	Aqueous medium (activated sludge), 20–25 °C	Laboratory	O_2 demand (ISO 14851)/ CO_2 evolution (ISO 14852)	**Test Validity**: Reference must reach > 60% biodegradation (14 days) **Pass Criteria (per BS EN 14987)**: > 90% ultimate biodegradation confirmed when indicator (O_2 or CO_2) approaches theoretical maximum (ThOD/ThCO_2)	[320–322]

(continued)

Table 3 (continued)

Test method/ standard	Environment + conditions	Test type	Biodegradable indicators	Test validity and criteria	References
OECD 301F	Aqueous, manometric respirometry	Laboratory	O_2 consumption (ThOD)	**Test Validity**: Reference > 60% (14 days) **Pass Criteria (Ready Biodegradability)**: ≥ 60% of ThOD within a 10-day window	[75]
DIN SPEC 4872	Aqueous, wastewater	Laboratory	Fiber release, biodegradation, ecotoxicity	**Pass Criteria**: Classifies textiles by microfiber release rate and % biodegradation (typically > 60%) in wastewater and ecotoxicological harmlessness (no single pass/fail)	[323]
ASTM D6691	Marine seawater, 30 ± 2 °C	Laboratory	CO_2 evolution	**Test Validity**: Reference must degrade > 70% **Pass Criteria (per ISO 22403)**: ≈90% of organic carbon converted to CO_2 within 6 months	[324, 325]
ISO 19679/ ISO 18830	Marine sediment–seawater interface, 15–25 °C	Laboratory	CO_2 evolution (ISO 19679)/O_2 demand (ISO 18830)	**Test Validity**: Reference must reach > 60% biodegradation **Pass Criteria (per ISO 22403)**: ≈90% of reference CO_2 evolution or O_2 consumption within 6–24 months	[325–327]
ISO 18830	Marine sediment–seawater interface, 15–25 °C	Laboratory	O_2 demand	**Test Validity**: Reference must reach > 60% biodegradation **Pass Criteria (per ISO 22403):** ≈90% O_2 consumption relative to reference	[325, 327]
ISO 22404	Marine surface / pelagic water, 15–25 °C	Laboratory	O_2 demand	**Test Validity**: Reference material must reach > 60% biodegradation **Pass Criteria (per ISO 22403)**: > 90% mineralization (O_2 uptake) for floating/suspended materials within 6 months	[325, 328]

(continued)

Table 3 (continued)

Test method/standard	Environment + conditions	Test type	Biodegradable indicators	Test validity and criteria	References
ISO 15985/ASTM D5511	High-solids anaerobic digestion, 37 °C (meso) or 52 °C (thermo)	Laboratory	Biogas (CH_4 + CO_2)	**Test Validity**: Reference must achieve > 70% carbon conversion (15–30 days); methods are technically equivalent **Pass Criteria**: No global specification; biogas plants or schemes use plant-specific thresholds, typically 50–70%	[329, 330]
ASTM D5526	Accelerated landfill conditions (anaerobic)	Laboratory	Biogas production	**Pass Criteria**: Long-term simulation (> 300 days); higher biogas yields indicate degradation potential (comparative assessment; no universal threshold)	[331]
ISO 14853/ISO 13975	Anaerobic aqueous medium; with sludge, ≈35 °C (ISO 14853)/general conditions (ISO 13975)	Laboratory	Biogas (CH_4 + CO_2)	**Pass Criteria**: Biodegradation confirmed when biogas production clearly exceeds background and approaches a readily degradable reference; no universal pass/fail threshold	[332, 333]

Note RH, relative humidity; $ThCO_2$, theoretical carbon dioxide; ThOD, theoretical oxygen demand; WHC, water-holding capacity standards such as EN 13432 and ASTM D6400 [48, 74, 130, 306, 307]. Complementary composting tests evaluate physical disintegration rather than mineralization: ISO 20200 and ISO 16929 assess fragmentation by sieve analysis and serve as screening tools within certification frameworks, while ISO/CD 17952 is being developed to address textile-specific performance indicators such as surface-area reduction and mechanical integrity loss [308, 309, 311]. Home compostability standards (AS 5810, NF T51-800) apply extended timeframes and mesophilic conditions, combining mineralization and disintegration criteria for materials intended for decentralized composting systems [312, 313].

Soil Biodegradation and Soil-Contact Testing

Soil biodegradation methodologies assess the ultimate aerobic biodegradation of materials in soil matrices and are particularly relevant for textiles intended for prolonged ground contact. Standards such as ISO 17556 and ASTM D5988 quantify biodegradation via oxygen demand or CO_2 evolution relative to theoretical carbon conversion, with explicit validity requirements based on reference material performance. Product-specific specifications, including EN 17033, extend these methods by defining pass criteria for agricultural textiles, linking biodegradation performance to functional durability [314–316]. In contrast, soil-burial strength-loss tests (ISO 11721-1/2, AATCC 30) are comparative methods used to evaluate rot susceptibility or rot resistance; they rely on relative changes in tensile or breaking strength and are therefore unsuitable for biodegradability certification but valuable for durability assessment in service environments [317–319].

Aquatic Biodegradability Test Methodologies

Aquatic biodegradation tests quantify mineralization in freshwater, wastewater, or marine environments using oxygen consumption or CO_2 evolution as primary indicators. ISO 14851 and ISO 14852 employ respirometry systems inoculated with activated sludge, while OECD 301F defines criteria for ready biodegradability based on oxygen demand within a defined time window [75, 320–322]. For wastewater-relevant applications, DIN SPEC 4872 extends assessment beyond mineralization to include microfiber release and ecotoxicological screening, providing a broader environmental performance classification rather than a binary pass/fail outcome [323]. Marine biodegradability is evaluated using seawater and sediment–water interface tests defined in ASTM D6691, ISO 18830, ISO 19679, and ISO 22404. These methods underpin marine biodegradability claims under ISO 22403, which specifies conversion thresholds relative to reference materials across different marine exposure scenarios [324–328].

Anaerobic Biodegradation and Landfill Simulation

Anaerobic biodegradation methodologies simulate oxygen-limited disposal routes, such as landfills or anaerobic digestion systems, by quantifying biogas production. Standards such as ISO 15985 and ASTM D5511 define high-solids anaerobic digestion tests, with test validity confirmed by the rapid conversion of a reference material [329, 330]. Long-term landfill simulation methods, such as ASTM D5526, provide comparative assessments of biogas yield over extended durations but do not define universal pass criteria, while aqueous anaerobic methods (ISO 14853; ISO 13975) confirm biodegradation via gas-evolution trends rather than fixed thresholds, reflecting the absence of globally harmonized certification requirements for landfill biodegradability [331–333].

Field Tests Under Natural Conditions

Field tests under natural conditions involve direct exposure of materials to real soil or aquatic environments, offering high ecological relevance but limited experimental control. They lack standardized control over key parameters such as microbial community composition, temperature and moisture profiles, and local geochemistry, and are therefore unsuitable as primary methods for regulatory compliance or certification. Spatial heterogeneity, seasonal fluctuations, and site-specific conditions introduce substantial variance, while fragmentation and sample migration complicate real-time monitoring of degradation mechanisms. Consequently, field studies are primarily used to validate trends observed in standardized laboratory or simulation tests, rather than to generate definitive biodegradability claims [6, 41].

Weathering Test

Weathering tests are not biodegradation methods but rather complementary preconditioning tools used to simulate long-term outdoor exposure of technical textiles prior to applying composting, soil, aquatic, or anaerobic biodegradation tests. They are therefore integrated into biodegradation assessment frameworks to ensure that materials are evaluated in an aged state that reflects realistic service conditions rather than as pristine specimens [15, 335]. They quantify physical and chemical degradation via color change, surface erosion, and loss of tensile or bursting strength, without measuring CO_2, O_2, or biogas evolution; ultimate biodegradation must therefore be assessed subsequently using appropriate biodegradation standards. Protocols such as ASTM D4329, ASTM G154, AATCC 186, AATCC TM111, ISO 877–1/2/3, and ISO 4892–3, therefore, occupy a complementary role in the testing framework, providing rigorous exposure and damage rating schemes that feed into, but do not replace, standardized biodegradability assessments [336–341].

5.3 *Challenges of Assessment*

Despite the availability of established conceptual frameworks and standardized methodologies for biodegradation testing, their practical application to technical textiles poses significant challenges. The following subsections address two interconnected issues: the standardization gaps in testing protocols that limit the applicability of existing standards to engineered textiles, and the performance–biodegradability trade-offs that arise when attempting to design materials that are simultaneously durable in service yet degradable at end-of-life [15, 197].

5.3.1 Standardization Gaps in Testing

A principal barrier to advancing biodegradable technical textiles is the absence of standardized testing protocols that accurately reflect the unique physicochemical properties of textile materials. The current regulatory framework predominantly relies on standards derived from the plastics industry, which do not capture the hierarchical structure and multiphase degradation behavior of textiles, each exhibiting distinct kinetics under different environmental conditions [15, 36, 197, 289]. Traditional textile microbiological tests, such as BS EN 14119 and AATCC 30, quantify resistance to rot, mildew, and microbial attack rather than promoting or measuring complete biodegradation, leaving a methodological gap for evaluating technic [5, 319, 342].

Within this framework, many studies still rely on gravimetric weight loss, tensile strength reduction, or visual surface changes as primary indicators of "biodegradation", which mainly capture physical disintegration and embrittlement rather than mineralization of polymer-derived carbon to CO_2 and biomass [20, 83, 343]. Without complementary respirometry endpoints, materials may pass disintegration-based criteria while continuing to generate persistent microplastics, creating a risk of hidden pollution despite nominal biodegradable claims [303, 304].

Standardized tests also employ optimized thermophilic or mesophilic conditions, controlled moisture, and selected inocula to accelerate degradation, which can substantially overestimate rates relative to cooler, nutrient-limited freshwater, marine, or soil environments where textile materials often accumulate [5, 19, 90]. Studies report substantially slower conversion and greater variability in freshwater, marine, and soil matrices than under industrial composting conditions, with biodegradation rates in unbuffered natural environments reported to be up to 42% lower than those observed in controlled laboratory settings, making it challenging to extrapolate certification results to real environmental persistence [15, 62, 85].

Standardized biodegradation tests also impose fixed exposure periods that are often much shorter than the service life and realistic degradation timescales of textiles, which can span several years in use and in the environment. Many tests require high mineralization within these short, fixed windows (e.g., $\geq 60\%$ in 28 days for ready biodegradability and $\geq 90\%$ in 6 months for industrial compostability specifications, as discussed in Sect. 5.1.1), and materials must meet these thresholds to be classified as biodegradable [47, 48, 75, 130, 344]. This framework compresses continuous biodegradation behavior into binary pass/fail outcomes, so slow-degrading but ultimately biodegradable textiles may be labelled "non-biodegradable" not because they cannot mineralize, but because their degradation kinetics are incompatible with the test timeframes, even when they would degrade over realistic multi-year environmental timescales [41, 279, 345, 346].

At the same time, fiber blends, high-performance finishes, and other functional additives common in technical textiles can inhibit microbial colonization or slow hydrolysis, yet these effects are only partially captured in current test protocols [15, 347]. In multi-component systems, one fiber or coating may mineralize while another persists as microplastic, so current standards struggle to classify partial biodegradation and to relate disintegration to genuine ecological safety [5, 49, 348]. Ultimately, the lack of harmonized, textile-specific biodegradability criteria and consistent test durations (often ranging from only 21 to 112 days) makes it challenging to define a clear threshold for "textile biodegradability" across freshwater, marine, and soil environments, so comparative assessment of durability and degradation performance remains scientifically tenuous [15, 62, 85].

5.3.2 Performance–Biodegradability Trade-Offs

A central obstacle to implementing biodegradable technical textiles is the fundamental inverse law of materials science: the physicochemical attributes required for high-performance service are intrinsic antagonists of biological degradation [25, 27, 62]. The very design strategies that deliver high performance and functional properties are achieved by engineering out the mechanisms of hydrolysis, oxidation, and microbial colonization on which biodegradation relies. This conflict manifests across structural, chemical, and temporal scales, effectively rendering many high-performance bio-based concepts theoretically attractive but practically difficult to realize [15, 44]. At the structural level, the functionality of sectors such as Geotech, Protech, and Indutech relies on extreme molecular order. Applications in ballistics, high-load filtration, or soil stabilization typically require polymer backbones with crystallinity fractions of approximately 60–80% and high chain orientation to achieve tensile strengths in the 800–1100 MPa range [349–351]. This dense architectural packing creates a physical occlusion effect, where the free volume between polymer chains is insufficient to accommodate infiltration of extracellular depolymerases, so even chemically hydrolysable or oxidatively labile backbones are largely confined to slow surface erosion rather than bulk depolymerization [44, 296, 299].

Beyond molecular architecture, the extensive use of functional finishes introduces a secondary layer of chemical antagonism, particularly in Agrotech, Hometech, Packtech, and Clothtech. These textiles function as composites in which the surface chemistry is deliberately designed to repel the environmental agents of decay. Agrotextiles depend on hindered amine light stabilizers (HALS) to endure solar radiation; however, these stabilizers actively quench the photo-oxidative radical formation required to initiate fragmentation of polyolefins [189, 254, 352]. Similarly, hydrophobic coatings in packaging and cross-linked adhesive systems in clothing prevent the water uptake necessary for hydrolytic enzyme activity, locking polymers into non-degradable or slowly degradable networks [353, 354]. Flame-retardant systems, particularly halogenated and some phosphorus-based chemistries used in Hometech and Protech, further complicate end-of-life by introducing persistent or toxic additives that can inhibit compost microflora and cause treated textiles to fail ecotoxicity criteria, even when the base fiber is nominally biodegradable [15, 355]. A more acute conflict exists in home and medical textiles, where antimicrobial finishes (e.g., AgNPs) are employed to ensure hygiene but inadvertently suppress the composting microflora, reducing the biological metabolism required for mineralization [15, 20, 83]. A temporal incompatibility further constrains sectors in which service life must substantially exceed the onset of degradation, such as Mobiltech, Buildtech, Meditech, and Sporttech. In automotive and construction applications, safety standards demand zero-creep tolerance and property retention over decades, making the molecular mobility and chain scission associated with biodegradation an unacceptable failure mode [350, 356]. Even in sport and medical applications, the requirements for high elasticity or viral impermeability necessitate thermoset-like cross-linking or multilayer lamination, structural forms that prevent the physical separation and solubility [357–361]. Finally, Oekotech presents a unique ecotoxicological paradox: while filtration and sorbent textiles are designed to trap pollutants such as heavy metals and hydrocarbons, biodegradation of the polymer matrix would re-release these concentrated toxins into the environment, causing the material to fail standard ecotoxicity assessments [362, 363]. Thus, across all twelve classes of technical textiles, the sector faces the challenge of developing "triggerable" degradation mechanisms that can bridge the gap between decades of required inertness and the rapid, environmentally safe decomposition demanded by circular-economy frameworks [27, 364, 365].

6 Conclusion

Integrating biodegradability into technical textiles addresses the challenge of balancing durability with environmental responsibility. As shown in this review, biodegradability depends on system factors, not just material type, highlighting a trade-off between stability and breakdown. The adoption of bio-based polymers reduces reliance on fossil fuels, but a renewable source does not guarantee the material will biodegrade at end-of-life.

At present, three fundamental barriers limit the widespread realization of biodegradable high-performance textiles. The first is a structural barrier, in which the high crystallinity and molecular orientation required for load-bearing applications physically restrict enzymatic access and depolymerization. The second is a chemical barrier, arising from the widespread use of protective finishes, such as antimicrobial agents, flame-retardant coatings, hydrophobic finishes, and UV stabilizers, that create an inhibitory microenvironment and actively suppress microbial colonization. Together, these constraints reinforce the durability–degradability paradox, whereby the physicochemical modifications essential for service performance simultaneously slow degradation kinetics and promote long-term environmental persistence. The third, and most critical, limitation is the assessment gap. Current reliance on plastics-based standards creates a false equivalence between industrial compostability and true environmental biodegradability. As demonstrated in this review, materials that perform well under thermophilic, managed composting conditions often persist in mesophilic soils or oligotrophic aquatic environments. In addition, many existing test protocols employ optimized laboratory conditions, short testing periods, and binary pass/fail criteria that fail to capture the complex, multi-phase degradation behavior of engineered textile structures. Consequently, indirect indicators such as weight loss or surface deterioration cannot be interpreted as evidence of actual biodegradability unless supported by direct measurements of mineralization under relevant environmental conditions.

Advancing sustainable technical textiles requires more than material substitution; it needs system-level design for end-of-life. Progress should emphasize fiber–finish systems with triggered-degradation mechanisms that remain stable during use but break down quickly upon disposal. Consistent, textile-specific testing protocols reflecting real-world environments are crucial for reliable assessment. Technical textiles must balance functional performance with the ability to re-enter biogeochemical cycles, supporting both innovation and ecological integrity.

References

1. Paul R (2019) High performance technical textiles: an overview. In: Paul R (ed) High performance technical textiles, ch 1. Wiley, pp 1–10. https://doi.org/10.1002/9781119325062.ch1
2. Byrne, C (2000) Technical textiles market—an overview. In: Horrocks AR, Anand SC (eds) Handbook of technical textiles, 2nd ed, ch 1. Woodhead Publishing, pp 1–23. https://doi.org/10.1533/9781855738966.1
3. McCarthy BJ (2016) An overview of the technical textiles sector. In: Horrocks AR, Anand SC (eds) Handbook of technical textiles: technical textile processes, 2nd ed, vol 1, ch 1. Woodhead Publishing, pp 1–20. https://doi.org/10.1016/B978-1-78242-458-1.00001-7.
4. Slater K (2000) Textiles and the environment. In: Horrocks AR, Anand SC (eds) Handbook of technical textiles, ch 19. Woodhead Publishing, pp 529–542. https://doi.org/10.1533/9781855738966.529.
5. Egan J, Salmon S (2022) Strategies and progress in synthetic textile fiber biodegradability. SN Appl Sci 4(1). https://doi.org/10.1007/s42452-021-04851-7
6. Rana S, Pichandi S, Parveen S, Fangueiro R (2014) Biodegradation studies of textiles and clothing products. In: Muthu SS (ed) Roadmap to sustainable textiles and clothing, textile science and clothing technology, 1st ed, ch 4. Springer, Singapore, pp 83–123. https://doi.org/10.1007/978-981-287-110-7_4
7. MarketsandMarkets (2025) Technical textiles market, industry size forecast. Available: https://www.marketsandmarkets.com/
8. Research and Markets (2025) Technical textiles global market report 2025. Available: https://www.researchandmarkets.com/
9. DeStefano V, Khan S, Tabada A (2020) Applications of PLA in modern medicine. Eng Regen 1:76. https://doi.org/10.1016/J.ENGREG.2020.08.002
10. Arshad K, Skrifvars M, Vivod V, Valh JV, Vončina B (2014) Biodegradation of natural textile materials in soil. Tekstilec 57(2):118–132. https://doi.org/10.14502/Tekstilec2014.57.118-132
11. Meng C (2025) Electrospinning PLLA/PCL blend fibre-based materials and their biomedical application: a mini review. Polymers (Basel) 17(20):2802. https://doi.org/10.3390/POLYM17202802
12. Ding H, Xie H, Singh S, Singh DN, He S, Wang J (2026) Geosynthetics and their applications. In: Xie H, Yan H, Wang Q (eds) Application of geosynthetics in waste containment ch 1. Elsevier, pp 1–37. https://doi.org/10.1016/B978-0-443-24852-8.00001-0

13. Rawal A, Shah T, Anand S (2010) Geotextiles: production, properties and performance. Text Prog 42(3):181–226. https://doi.org/10.1080/00405160903509803
14. Sabir T (2017) Fibers used for high-performance apparel. In: J Mcloughlin J, Sabir T (eds) High-performance apparel: materials, development, and applications, ch 2. Woodhead Publishing, pp 7–32. https://doi.org/10.1016/B978-0-08-100904-8.00002-X.
15. Skilbeck OJ, Blackburn RS, Kay P (2025) A review on the biodegradation of textiles in the environment. Environ Toxicol Chem.https://doi.org/10.1093/etojnl/vgaf229
16. Mohanan N, Montazer Z, Sharma PK, Levin DB (2020) Microbial and enzymatic degradation of synthetic plastics. Front Microbiol 11:580709. https://doi.org/10.3389/FMICB.2020.580709
17. Textile Exchange (2025) Materials market report. Available: https://textileexchange.org/
18. Tokiwa Y, Calabia BP (2007) Biodegradability and biodegradation of polyesters. J Polym Environ 15(4):259–267. https://doi.org/10.1007/S10924-007-0066-3
19. Erdal NB, Hakkarainen M (2022) Degradation of cellulose derivatives in laboratory, man-made, and natural environments. Am Chem Soc 11. https://doi.org/10.1021/acs.biomac.2c00336.
20. Lykaki M et al (2021) The influence of textile finishing agents on the biodegradability of shed fibres. Green Chem 23(14):5212–5221. https://doi.org/10.1039/d1gc00883h
21. Zambrano MC, Pawlak JJ, Daystar J, Ankeny M, Venditti RA (2021) Impact of dyes and finishes on the aquatic biodegradability of cotton textile fibers and microfibers released on laundering clothes: correlations between enzyme adsorption and activity and biodegradation rates. Mar Pollut Bull 165. https://doi.org/10.1016/j.marpolbul.2021.112030
22. Yıldırım S (2025) Plastic biodegradation: potential of microbial enzymes. GISTU J Health Biol Sci 57–78. https://doi.org/10.65308/GJOHBS.2025.010
23. Deroiné M et al (2014) Accelerated ageing of polylactide in aqueous environments: comparative study between distilled water and seawater. Polym Degrad Stab 108:319–329. https://doi.org/10.1016/j.polymdegradstab.2014.01.020
24. Pavko Čuden A (2023) Sustainability in functional and technical textiles. In: Functional and technical textiles. Elsevier, pp 779–818. https://doi.org/10.1016/B978-0-323-91593-9.00012-2
25. Aarsen CV, Liguori A, Mattsson R, Sipponen MH, Hakkarainen M (2024) Designed to degrade: tailoring polyesters for circularity. Chem Rev 124(13):8473–8515. https://doi.org/10.1021/acs.chemrev.4c00032
26. Patti A, Acierno D (2022) Towards the sustainability of the plastic industry through biopolymers: properties and potential applications to the textiles world. MDPI. https://doi.org/10.3390/polym14040692
27. Withana PA et al (2025) Biodegradable plastics in soils: sources, degradation, and effects. R Soc Chem. https://doi.org/10.1039/d4em00754a
28. Fiandra EF, Shaw L, Starck M, McGurk CJ, Mahon CS (2023) Designing biodegradable alternatives to commodity polymers. Chem Soc Rev 52(23):8085–8105. https://doi.org/10.1039/D3CS00556A
29. Suh S et al (2024) Conceptual framework for identifying polymers of concern. Front Sustain 5:1399431. https://doi.org/10.3389/FRSUS.2024.1399431/FULL
30. Caicedo C, Melo-López L, Cabello-Alvarado C, Cruz-Delgado VJ, Ávila-Orta CA (2019) Biodegradable polymer nanocomposites applied to technical textiles: a review. DYNA (Colombia) 86(211):288–299. https://doi.org/10.15446/DYNA.V86N211.80230
31. Patwary S (2020) Clothing and textile sustainability: current state of environmental challenges and the ways forward. Textile Leather Rev 3(3):158–173. https://doi.org/10.31881/TLR.2020.16
32. An Y, Padermshoke A, Van Nguyen T, Takahara A (2024) Surface chemistry in environmental degradation of polymeric solids. Langmuir 40(18):9336–9344. https://doi.org/10.1021/ACS.LANGMUIR.4C00731
33. Kliem S, Kreutzbruck M, Bonten C (2020) Review on the biological degradation of polymers in various environments. Materials 13(20):4586. https://doi.org/10.3390/MA13204586

34. Muniyasamy S, Ofosu O, John MJ, Anandjiwala RD (2016) Mineralization of Poly(lactic acid) (PLA), Poly(3-hydroxybutyrate-co-valerate) (PHBV) and PLA/PHBV blend in compost and soil environments. J Renew Mater 4(2):133–145. https://doi.org/10.7569/JRM.2016.634104
35. Song JH, Murphy RJ, Narayan R, Davies GBH (2009) Biodegradable and compostable alternatives to conventional plastics. Philos Trans R Soc B: Biol Sci 364(1526):2127–2139. https://doi.org/10.1098/RSTB.2008.0289
36. Pinto N, Amorim Â, Marinho A, Vieira de Castro P (2025) Textile degradation: a comprehensive review and experimental approach. J Textile Eng Fash Technol 11(5). https://doi.org/10.15406/JTEFT.2025.11.00429
37. Tantawi O et al (2025) Designing for degradation: the importance of considering biotic and abiotic polymer degradation. Environ Sci Process Impacts 27(5):1303–1316. https://doi.org/10.1039/D5EM00079C
38. Russell JR et al (2011) Biodegradation of polyester polyurethane by endophytic fungi. Appl Environ Microbiol 77(17):6076–6084. https://doi.org/10.1128/AEM.00521-11
39. Negi A (2025) Environmental impact of textile materials: challenges in fiber-dye chemistry and implication of microbial biodegradation. Polymers (Basel) 17(7):871. https://doi.org/10.3390/POLYM17070871
40. Okolie O et al (2023) Bio-based sustainable polymers and materials: from processing to biodegradation. J Compos Sci 7:213. https://doi.org/10.3390/JCS7060213
41. Pires JRA, Souza VGL, Fuciños P, Pastrana L, Fernando AL (2022) Methodologies to assess the biodegradability of bio-based polymers—current knowledge and existing gaps. MDPI. https://doi.org/10.3390/polym14071359
42. Samir A, Ashour FH, Hakim AAA, Bassyouni M (2022) Recent advances in biodegradable polymers for sustainable applications. NPJ Mater Degrad 6(1):68–. https://doi.org/10.1038/s41529-022-00277-7
43. Coltelli MB, Morganti P, Lazzeri A (2023) Sustainability assessment, environmental impact, and recycling strategies of biodegradable polymer nanocomposites. In: Biodegradable and biocompatible polymer nanocomposites: processing, characterization, and applications. Elsevier, pp 699–737. https://doi.org/10.1016/B978-0-323-91696-7.00016-7
44. Tokiwa Y, Calabia BP, Ugwu CU, Aiba S (2009) Biodegradability of plastics. Int J Mol Sci 10(9):3722–3742. https://doi.org/10.3390/ijms10093722
45. Wackett LP (2019) Bio-based and biodegradable plastics: an annotated selection of World Wide Web sites relevant to the topics in microbial biotechnology. Microb Biotechnol 12(6):1492–1493. https://doi.org/10.1111/1751-7915.13502
46. Park CH, Kang YK, Im SS (2004) Biodegradability of cellulose fabrics. J Appl Polym Sci 94(1):248–253. https://doi.org/10.1002/app.20879
47. ISO (2021) ISO 17088:2021—Plastics—Organic recycling—Specifications for compostable plastics. Available: https://www.iso.org/home.html
48. BSI (2000) BS EN 13432:2000 packaging. Requirements for packaging recoverable through composting and biodegradation. Test scheme and evaluation criteria for the final acceptance of packaging. Available: https://www.bsigroup.com/
49. Mehta S (2023) Biodegradable textile polymers: a review of current scenario and future opportunities. Taylor and Francis Ltd. https://doi.org/10.1080/21622515.2023.2227391.
50. Lucas N, Bienaime C, Belloy C, Queneudec M, Silvestre F, Nava-Saucedo JE (2008) Polymer biodegradation: mechanisms and estimation techniques—a review. https://doi.org/10.1016/j.chemosphere.2008.06.064
51. Tripa S, García PD, Indrie L, Mikucioniene D (2024) Solutions to reduce the environmental pressure exerted by technical textiles: a review. Industria Textila 75(1):66–74. https://doi.org/10.35530/IT.075.01.202367
52. Luján-Ornelas C, Güereca LP, Franco-García ML, Heldeweg M (2020) A life cycle thinking approach to analyse sustainability in the textile industry: a literature review. Sustainability 12(23):10193. https://doi.org/10.3390/SU122310193
53. Emadian SM, Onay TT, Demirel B (2017) Biodegradation of bioplastics in natural environments. Waste Manage 59:526–536. https://doi.org/10.1016/j.wasman.2016.10.006

54. Gewert B, Plassmann MM, Macleod M (2015) Pathways for degradation of plastic polymers floating in the marine environment. Environ Sci Process Impacts 17(9):1513–1521. https://doi.org/10.1039/C5EM00207A
55. Bher A, Mayekar PC, Auras RA, Schvezov CE (2022) Biodegradation of biodegradable polymers in mesophilic aerobic environments. MDPI. https://doi.org/10.3390/ijms232012165
56. Silva RRA, Marques CS, Arruda TR, Teixeira SC, de Oliveira TV (2023) Biodegradation of polymers: stages, measurement, standards and prospects. Multidisciplinary Digital Publishing Institute (MDPI). https://doi.org/10.3390/macromol3020023
57. Flemming HC, Wingender J (2010) The biofilm matrix. Nat Rev Microbiol 8(9):623–633. https://doi.org/10.1038/NRMICRO2415
58. Ceretti DVA, Edeleva M, Cardon L, D'hooge DR (2023) Molecular pathways for polymer degradation during conventional processing, additive manufacturing, and mechanical recycling. Molecules 28(5):2344. https://doi.org/10.3390/MOLECULES28052344
59. Woodard LN, Grunlan MA (2018) Hydrolytic degradation and erosion of polyester biomaterials. ACS Macro Lett 7(8):976. https://doi.org/10.1021/ACSMACROLETT.8B00424
60. Møllebjerg A, Palmén LG, Gori K, Meyer RL (2021) The bacterial life cycle in textiles is governed by fiber hydrophobicity. Microbiol Spectr 9(2):e01185-e1221. https://doi.org/10.1128/SPECTRUM.01185-21
61. Szostak-Kotowa J (2004) Biodeterioration of textiles. In: International biodeterioration and biodegradation, pp 165–170. https://doi.org/10.1016/S0964-8305(03)00090-8
62. Zambrano MC, Pawlak JJ, Venditti RA (2020) Effects of chemical and morphological structure on biodegradability of fibers, fabrics, and other polymeric materials. BioResources 15(4):9786–9833. https://doi.org/10.15376/biores.15.4.Zambrano
63. Chavez-Linares P, Hoppe S, Chevalot I (2025) Recycling and degradation pathways of synthetic textile fibers such as polyamide and elastane. Global Chall 9(4):2400163. https://doi.org/10.1002/GCH2.202400163
64. Zeenat AE, Bukhari DA, Shamim S, Rehman A (2021) Plastics degradation by microbes: a sustainable approach. J King Saud Univ Sci 33(6):101538. https://doi.org/10.1016/J.JKSUS.2021.101538
65. Souza GD et al (2023) Interspecies interactions determine growth dynamics of biopolymer-degrading populations in microbial communities. Proc Natl Acad Sci USA 120(44):e2305198120. https://doi.org/10.1073/PNAS.2305198120
66. Iwańczuk A, Kozłowski M, Łukaszewicz M, Jabłoński S (2014) Anaerobic biodegradation of polymer composites filled with natural fibers. J Polym Environ 23(2):277–282. https://doi.org/10.1007/S10924-014-0690-7
67. Falzarano M, Polettini A, Pomi R, Rossi A, Zonfa T (2023) Anaerobic biodegradability of commercial bioplastic products: systematic bibliographic analysis and critical assessment of the latest advances. Materials 16(6):2216. https://doi.org/10.3390/MA16062216
68. Quecholac-Piña X, Hernández-Berriel MDC, Mañón-Salas MDC, Espinosa-Valdemar RM, Vázquez-Morillas A (2020) Degradation of plastics under anaerobic conditions: a short review. Polymers (Basel) 12(1):109. https://doi.org/10.3390/POLYM12010109
69. García-Depraect O et al (2022) Biodegradation of bioplastics under aerobic and anaerobic aqueous conditions: Kinetics, carbon fate and particle size effect. Bioresour Technol 344:126265. https://doi.org/10.1016/J.BIORTECH.2021.126265
70. Innocenti FD (2003) Biodegradability and compostability. In: Chiellini E, Solaro R (eds) Biodegradable polymers and plastics, Pisa, Italy, pp 33–45. https://doi.org/10.1007/978-1-4419-9240-6_2
71. Coltelli MB, Gigante V, Aliotta L, Lazzeri A (2024) Recyclability perspectives of the most diffused biobased and biodegradable plastic materials. Macromol 4(2):401–419. https://doi.org/10.3390/MACROMOL4020023
72. Reuschenbach P, Pagga U, Strotmann U (2003) A critical comparison of respirometric biodegradation tests based on OECD 301 and related test methods. Water Res 37(7):1571–1582. https://doi.org/10.1016/S0043-1354(02)00528-6

73. Strotmann U, Reuschenbach P, Schwarz H, Pagga U (2004) Development and evaluation of an online CO_2 evolution test and a multicomponent biodegradation test system. Appl Environ Microbiol 70(8):4621. https://doi.org/10.1128/AEM.70.8.4621-4628.2004
74. ISO (2012) ISO 14855-1:2012—determination of the ultimate aerobic biodegradability of plastic materials under controlled composting conditions—method by analysis of evolved carbon dioxide—part 1: general method. Available: https://www.iso.org/home.html
75. OECD (1992) Test No. 301: Ready Biodegradability. https://doi.org/10.1787/9789264070349-en
76. Folino A, Karageorgiou A, Calabrò PS, Komilis D (2020) Biodegradation of wasted bioplastics in natural and industrial environments: a review. Sustainability 12(15):6030. https://doi.org/10.3390/SU12156030
77. Sikorska W, Musioł M, Zawidlak-Węgrzyńska B, Rydz J (2021) End-of-life options for (Bio)degradable polymers in the circular economy. Adv Polym Technol 2021(1):6695140. https://doi.org/10.1155/2021/6695140
78. Rong-or C, Pongputthipat W, Ruksakulpiwat Y, Chumsamrong P (2024) Soil burial degradation of bio-composite films from poly(lactic acid), natural rubber, and rice straw. Polym Bull 81(12):10729–10746. https://doi.org/10.1007/S00289-024-05229-6
79. Xin J, Hu J, Song Z, Zhao C, Chen D, Yang W (2025) Biodegradation and ecotoxicity of Poly(1-Hexene-Maleic Anhydride): insights into environmental behavior and biological impact. ACS Appl Mater Interfaces 17(44):60522–60533. https://doi.org/10.1021/ACSAMI.5C14178
80. Smagin AV, Sadovnikova NB, Vasenev VI, Smagina MV (2018) Biodegradation of some organic materials in soils and soil constructions: experiments, modeling and prevention. Materials 11(10):1889. https://doi.org/10.3390/MA11101889.
81. Kanakannavar S, Pitchaimani J, Thalla A, Rajesh M (2022) "Biodegradation properties and thermogravimetric analysis of 3D braided flax PLA textile composites. J Ind Textiles 51(1_suppl):1066S–1091S. https://doi.org/10.1177/15280837211010666
82. Gonzalez JM, Santana MM, Gomez EJ, Delgado JA (2023) Soil thermophiles and their extracellular enzymes: a set of capabilities able to provide significant services and risks. Microorganisms 11(7):1650. https://doi.org/10.3390/MICROORGANISMS11071650
83. Milošević M, Krkobabić A, Radoičić M, Šaponjić Z, Radetić T, Radetić M (2017) Biodegradation of cotton and cotton/polyester fabrics impregnated with Ag/TiO_2 nanoparticles in soil. Carbohydr Polym 158:77–84. https://doi.org/10.1016/j.carbpol.2016.12.006
84. Siddiquee KM, Maksud Helali M, Gafur MA, Chakraborty S (2014) Investigation of an optimum method of biodegradation process for jute polymer composites. Am J Eng Res (AJER) 03:200–206. Available: www.ajer.org
85. Podgornik BB, Šandric S, Kert M (2021) Microencapsulation for functional textile coatings with emphasis on biodegradability—a systematic review. Coatings 11(11). https://doi.org/10.3390/coatings11111371
86. Šerá J et al (2022) Biodegradable polyesters and low molecular weight polyethylene in soil: interrelations of material properties, soil organic matter substances, and microbial community. Int J Mol Sci 23(24):15976. https://doi.org/10.3390/IJMS232415976
87. Williams BM, Houseman GR (2014) Experimental evidence that soil heterogeneity enhances plant diversity during community assembly. J Plant Ecol 7(5):461–469. https://doi.org/10.1093/JPE/RTT056
88. Tomšič B, Klemenčič D, Simončič B, Orel B (2011) Influence of antimicrobial finishes on the biodeterioration of cotton and cotton/polyester fabrics: leaching versus bio-barrier formation. Polym Degrad Stab 96(7):1286–1296. https://doi.org/10.1016/j.polymdegradstab.2011.04.004
89. Tomšič B, Simončič B, Orel B, Vilčnik A, Spreizer H (2007) Biodegradability of cellulose fabric modified by imidazolidinone. Carbohydr Polym 69(3):478–488. https://doi.org/10.1016/j.carbpol.2007.01.003
90. Zambrano MC, Pawlak JJ, Daystar J, Ankeny M, Goller CC, Venditti VA (2020) Aerobic biodegradation in freshwater and marine environments of textile microfibers generated in

clothes laundering: effects of cellulose and polyester-based microfibers on the microbiome. Mar Pollut Bull 151. https://doi.org/10.1016/j.marpolbul.2019.110826
91. Wataniyakun W, Le Gall M, El Rakwe M, Karl CW, Larsen RB (2025) Biodegradable fishing gears: a potential solution to ghost fishing and marine plastic pollution. Mar Pollut Bull 212:117607. https://doi.org/10.1016/J.MARPOLBUL.2025.117607
92. Le Gué L, Davies P, Arhant M, Vincent B, Tanguy E (2023) Mitigating plastic pollution at sea: natural seawater degradation of a sustainable PBS/PBAT marine rope. Mar Pollut Bull 193:115216. https://doi.org/10.1016/J.MARPOLBUL.2023.115216
93. Vesamäki JS, Laine MB, Nissinen R, Taipale SJ (2024) Plastic and terrestrial organic matter degradation by the humic lake microbiome continues throughout the seasons. Environ Microbiol Rep 16(3):e13302. https://doi.org/10.1111/1758-2229.13302
94. Newton RJ, Jones SE, Eiler A, McMahon KD, Bertilsson S (2011) A guide to the natural history of freshwater lake bacteria. Microbiol Mol Biol Rev 75(1):14–49. https://doi.org/10.1128/MMBR.00028-10
95. Bocci V et al (2024) Freshwater plastisphere: a review on biodiversity, risks, and biodegradation potential with implications for the aquatic ecosystem health. Front Microbiol 15:1395401. https://doi.org/10.3389/FMICB.2024.1395401
96. Perez Calderon LJ et al (2019) Pressure and temperature effects on deep-sea hydrocarbon-degrading microbial communities in subarctic sediments. Microbiologyopen 8(6):1. https://doi.org/10.1002/MBO3.768
97. Asseri AA, Coello-Camba A, Agustí S (2025) Fungal planktonic community related to salinity and temperature in an oligotrophic sea. Front Microbiol 16:1435925. https://doi.org/10.3389/fmicb.2025.1435925
98. Price PB, Sowers T (2004) Temperature dependence of metabolic rates for microbial growth, maintenance, and survival. Proc Natl Acad Sci USA 101(13):4631–4636. https://doi.org/10.1073/PNAS.0400522101
99. Lavagnolo MC, Poli V, Zampini AM, Grossule V (2024) Biodegradability of bioplastics in different aquatic environments: a systematic review. J Environ Sci 142:169–181. https://doi.org/10.1016/J.JES.2023.06.013
100. Bagheri AR, Laforsch C, Greiner A, Agarwal S (2017) Fate of so-called biodegradable polymers in seawater and freshwater. Glob Challenges 1(4). https://doi.org/10.1002/GCH2.201700048
101. Chen H (2022) Biodegradable plastics in the marine environment: a potential source of risk? Water Emerg Contam Nanoplastics 1(4). https://doi.org/10.20517/WECN.2022.11
102. Omura T et al (2024) Microbial decomposition of biodegradable plastics on the deep-sea floor. Nat Commun 15(1):568–. https://doi.org/10.1038/s41467-023-44368-8
103. Hyun Y, Kwak HW, Jin HJ (2022) Biodegradation of aliphatic polyesters under composting and seawater conditions. Polymer (Korea) 46(3):318–326. https://doi.org/10.7317/PK.2022.46.3.318
104. Wang GX, Huang D, Ji JH, Völker C, Wurm FR (2020) Seawater-degradable polymers-fighting the marine plastic pollution. Adv Sci (Weinh) 8(1). https://doi.org/10.1002/ADVS.202001121
105. Eich A, Weber M, Lott C (2021) Disintegration half-life of biodegradable plastic films on different marine beach sediments. PeerJ 9. https://doi.org/10.7717/PEERJ.11981
106. Lott C et al (2021) Half-life of biodegradable plastics in the marine environment depends on material, habitat, and climate zone. Front Mar Sci 8:662074. https://doi.org/10.3389/FMARS.2021.662074/BIBTEX
107. Krause S, Molari M, Gorb EV, Gorb SN, Kossel E, Haeckel M (2020) Persistence of plastic debris and its colonization by bacterial communities after two decades on the abyssal seafloor. Sci Rep 10(1):9484. https://doi.org/10.1038/S41598-020-66361-7
108. Jacquin J et al (2019) Microbial ecotoxicology of marine plastic debris: a review on colonization and biodegradation by the 'plastisphere. Front Microbiol 10:865. https://doi.org/10.3389/FMICB.2019.00865

109. Nizamuddin S, Baloch AJ, Chen C, Arif M, Mubarak NM (2024) Bio-based plastics, biodegradable plastics, and compostable plastics: biodegradation mechanism, biodegradability standards and environmental stratagem. Int Biodeterior Biodegradation 195:105887. https://doi.org/10.1016/J.IBIOD.2024.105887
110. Bher A, Cho Y, Auras R (2023) Boosting degradation of biodegradable polymers. Macromol Rapid Commun 44(5):2200769. https://doi.org/10.1002/MARC.202200769
111. Eissenberger K et al (2023) Approaches in sustainable, biobased multilayer packaging solutions. Polymers (Basel) 15(5):1184. https://doi.org/10.3390/POLYM15051184
112. Fakhrizada H, Dahman Y (2025) Degradable alternatives to single-use plastics: mechanisms, materials, and strategies for sustainable polyolefin replacement. Molecules 30(21):4301. https://doi.org/10.3390/MOLECULES30214301
113. Finore I et al (2023) Thermophilic bacteria and their thermozymes in composting processes: a review. Chem Biol Technol Agric 10(1):7–. https://doi.org/10.1186/S40538-023-00381-Z
114. Sayara T, Sánchez A (2019) A review on anaerobic digestion of lignocellulosic wastes: pretreatments and operational conditions. Appl Sci 2019, 9(21):4655. https://doi.org/10.3390/APP9214655
115. Xavier CAN, Moset V, Wahid R, Møller HB (2015) The efficiency of shredded and briquetted wheat straw in anaerobic co-digestion with dairy cattle manure. Biosyst Eng 139:16–24. https://doi.org/10.1016/J.BIOSYSTEMSENG.2015.07.008
116. Tharmarajah N, Shahbaz K, Baroutian S (2024) Textile waste pretreatment for anaerobic digestion: a review and technology feasibility study. J Chem Technol Biotechnol 99(11):2213–2227. https://doi.org/10.1002/JCTB.7728
117. Barlaz MA, Green RB, Chanton JP, Goldsmith CD, Hater GR (2004) Evaluation of a biologically active cover for mitigation of landfill gas emissions. Environ Sci Technol 38(18):4891–4899. https://doi.org/10.1021/ES049605B
118. EPA. Basic information about landfill gas. Environmental Protection Agency, US. Available: https://www.epa.gov/
119. Krause MJ, Detwiler N, Eades W, Marro D, Schwarber A, Tolaymat T (2023) Understanding landfill gas behavior at elevated temperature landfills. Waste Manag 165:83. https://doi.org/10.1016/J.WASMAN.2023.04.023
120. Manjunatha GS, Chavan D, Lakshmikanthan P, Singh L, Kumar S, Kumar R (2020) Specific heat and thermal conductivity of municipal solid waste and its effect on landfill fires. Waste Manage 116:120–130. https://doi.org/10.1016/J.WASMAN.2020.07.033
121. Meyer-Dombard DR, Bogner JE, Malas J (2020) A review of landfill microbiology and ecology: a call for modernization with 'next generation' technology. Front Microbiol 11:1127. https://doi.org/10.3389/FMICB.2020.01127
122. Krause MJ, Eades W, Detwiler N, Marro D, Schwarber A, Tolaymat T (2023) Assessing moisture contributions from precipitation, waste, and leachate for active municipal solid waste landfills. J Environ Manage 344:118443. https://doi.org/10.1016/J.JENVMAN.2023.118443
123. Barlaz MA, Staley BF, De Los Reyes III FL (2010) Anaerobic biodegradation of solid waste. In: Mitchell R, Gu J-D (eds) Environmental microbiology, 2nd ed, ch 12. Wiley-Blackwell, Hoboken, New Jersey, pp 281–299. https://doi.org/10.1002/9780470495117.ch12
124. Harrison JP, Boardman C, O'Callaghan K, Delort AM, Song J (2018) Biodegradability standards for carrier bags and plastic films in aquatic environments: a critical review. R. Soc. Open Sci. 5(5):171792. https://doi.org/10.1098/RSOS.171792
125. Choe S, Kim Y, Won Y, Myung J (2021) Bridging three gaps in biodegradable plastics: misconceptions and truths about biodegradation. Front Chem 9:671750. https://doi.org/10.3389/FCHEM.2021.671750
126. Blanco I, Ingrao C, Siracusa V (2020) Life-cycle assessment in the polymeric sector: a comprehensive review of application experiences on the Italian scale. Polymers 12(6):1212. https://doi.org/10.3390/POLYM12061212
127. Kubowicz S, Booth AM (2017) Biodegradability of plastics: challenges and misconceptions. Environ Sci Technol 51(21):12058–12060. https://doi.org/10.1021/ACS.EST.7B04051

128. Wu J et al (2024) Biodegradation: the best solution to the world problem of discarded polymers. Bioresour Bioprocess 11(1):79–. https://doi.org/10.1186/S40643-024-00793-1
129. Dallaev R, Papež N, Allaham MM, Holcman V (2025) Biodegradable polymers: properties, applications, and environmental impact. Polymers 17(14):1981. https://doi.org/10.3390/POLYM17141981
130. ASTM International (2021) ASTM D6400-21, standard specification for labeling of plastics designed to be aerobically composted in municipal or industrial facilities. https://doi.org/10.1520/D6400-21
131. OECD (2014) Test No. 310: ready biodegradability—CO_2 in sealed vessels (headspace test). OECD Publishing. https://doi.org/10.1787/9789264224506-EN.
132. Nawab Y, Ahmad S (2020) Introduction. In: Ahmad S, Rasheed A, Nawab Y (eds) Fibers for technical textiles. Topics in mining, metallurgy and materials engineering, ch 1. Springer, Cham, pp 1–20. https://doi.org/10.1007/978-3-030-49224-3_1
133. Rasheed A (2020) Classification of technical textiles. In: Ahmad S, Rasheed A, Nawab Y (eds) Fibers for technical textiles. Topics in mining, metallurgy and materials engineering ch 3. Springer, Cham, pp 49–64. https://doi.org/10.1007/978-3-030-49224-3_3
134. Gautam S, Sharma D, Thakur A (2025) Agro-textiles for sustainable agriculture: materials, manufacturing, applications, and future prospects. Int J Adv Biochem Res 9(11S):656–660. https://doi.org/10.33545/26174693.2025.V9.I11SI.6298
135. Sharma N, Allardyce B, Rajkhowa R, Adholeya A, Agrawal R (2022) A substantial role of agro-textiles in agricultural applications. Front Plant Sci 13:895740. https://doi.org/10.3389/FPLS.2022.895740
136. Briassoulis D (2023) Agricultural plastics as a potential threat to food security, health, and environment through soil pollution by microplastics: problem definition. Sci Total Environ 892:164533. https://doi.org/10.1016/J.SCITOTENV.2023.164533
137. Friese D, Scheurer M, Hahn L, Gries T, Cherif C (2022) Textile reinforcement structures for concrete construction applications—a review. J Compos Mater 56(26):4041–4064. https://doi.org/10.1177/00219983221127181
138. Tabor J, Ghosh T (2019) Building and construction textiles. In: Paul R (ed) High performance technical textiles, ch 11. Wiley, pp 319–352. https://doi.org/10.1002/9781119325062.CH11
139. Ruckdashel RR, Khadse N, Park JH (2022) Smart E-textiles: overview of components and outlook. Sensors 22(16):6055. https://doi.org/10.3390/S22166055
140. Cho S, Chang T, Yu T, Lee CH (2022) Smart electronic textiles for wearable sensing and display. Biosensors (Basel) 12(4):222. https://doi.org/10.3390/BIOS12040222
141. Zhang Q, Kan CW (2018) A review of fusible interlinings usage in garment manufacture. Polymers 10(11):1230. https://doi.org/10.3390/POLYM10111230
142. Ahmad Z, Naeem MS, Jabbar A, Irfan M (2020) Fibers for other technical textiles applications. In: Ahmad S, Rasheed A, Nawab Y (eds) Fibers for technical textiles. Topics in mining, metallurgy and materials engineering, ch 10. Springer, Cham, pp 201–220. https://doi.org/10.1007/978-3-030-49224-3_10.
143. Rawal A, Shah TH, Anand SC (2016) Geotextiles in civil engineering. In: Horrocks AR, Anand SC (eds) Handbook of technical textiles: Technical textile applications, vol 2, ch 4. Woodhead Publishing, pp 111–133. https://doi.org/10.1016/B978-1-78242-465-9.00004-5
144. Hsieh CW (2016) Geotextiles in agriculture and aquaculture. In: Geotextiles: from design to applications, pp 511–530. https://doi.org/10.1016/B978-0-08-100221-6.00023-1.
145. Chai JC (2016) Geotextiles used in drainage. In: Koerner RM (ed) Geotextiles: from design to applications, ch 13. Woodhead Publishing, pp 277–303. https://doi.org/10.1016/B978-0-08-100221-6.00013-9
146. Koerner GR (2016) Geotextiles used in separation. In: Koerner RM (ed) Geotextiles: from design to applications, ch 11. Woodhead Publishing, pp 239–256. https://doi.org/10.1016/B978-0-08-100221-6.00011-5
147. Nistorac A, Bodoga A, Ailenei CE, Loghin M-C (2024) Sustainable horizons: exploring technical textiles and environmental responsibility. In: International symposium "technical textiles—present and future". Sciendo, pp 136–140. https://doi.org/10.2478/9788367405355-021

148. Belino N, Fangueiro R, Rana S, Glampedaki P, Priniotakis G (2019) Medical and healthcare textiles. In: Paul R (ed) High performance technical textiles, ch 4. Wiley, pp 69–105. https://doi.org/10.1002/9781119325062.ch4
149. Zhezhova S, Jordeva S, Golomeova-Longurova S, Jovanov S (2021) Application of technical textile in medicine. Tekstilna industrija 69(2):21–29. https://doi.org/10.5937/TEKSTIND2102021Z
150. Qin Q (2016) An overview of medical textile products. In: Qin Y (ed) Medical textile materials, ch 2. Elsevier, pp 13–22. https://doi.org/10.1016/B978-0-08-100618-4.00002-9
151. Horrocks AR (2016) Technical textiles in transport (land, sea, and air). In: Horrocks AR, Anand SC (eds) Handbook of technical textiles: technical textile applications, ch 11. Woodhead Publishing, pp 325–356. https://doi.org/10.1016/B978-1-78242-465-9.00011-2
152. Fung W, Hardcastle M (2001) Other textile applications. In: Fung W, Hardcastle M (eds) Textiles in automotive engineering, ch 7. Elsevier, pp 227–253. https://doi.org/10.1533/9781855738973.227
153. Chen X, Zhou Y (2016) Technical textiles for ballistic protection. In: Horrocks AR, Anand SC (Eds) Handbook of technical textiles: technical textile applications, 2nd ed., vol 2, ch 6. Woodhead Publishing, pp 169–192. https://doi.org/10.1016/B978-1-78242-465-9.00006-9
154. Horrocks AR (2016) Technical fibres for heat and flame protection. In: Horrocks AR, Anand SC (eds) Handbook of technical textiles: technical textile applications, vol. 2, ch 8. Woodhead Publishing, pp 237–270. https://doi.org/10.1016/B978-1-78242-465-9.00008-2
155. Crown EM, Batcheller JC (2016) Technical textiles for personal thermal protection. In: Horrocks AR, Anand SC (eds) Handbook of technical textiles: technical textile applications, 2nd ed., vol 2, ch 9. Woodhead Publishing, pp 271–285. https://doi.org/10.1016/B978-1-78242-465-9.00009-4
156. Holmes DA, Horrocks AR (2016) Technical textiles for survival. In: Horrocks AR, Anand SC (eds) Handbook of technical textiles: technical textile applications, 2nd ed., vol 2, ch 10. Woodhead Publishing, pp 287–323. https://doi.org/10.1016/B978-1-78242-465-9.00010-0
157. Ahmad S, Ullah T, Ziauddin (2020) Fibers for technical textiles. In: Ahmad S, Rasheed A, Nawab, Y (eds) Fibers for technical textiles. Topics in mining, metallurgy and materials engineering, ch 2. Springer, Cham, pp 21–47. https://doi.org/10.1007/978-3-030-49224-3_2
158. Menezes E (2000) Technical fibres. In: Horrocks AR, Anand SC (eds) Handbook of technical textiles, vol 43, no 8, ch 2. Woodhead Publishing, pp 24–41. https://doi.org/10.1533/9781855738966.24
159. Miraftab M (2016) Technical fibres: recent advances. In: Horrocks AR, Anand SC (eds) Handbook of technical textiles: technical textile processes, 2nd ed, vol 1, ch 2. Woodhead Publishing, pp 21–41. https://doi.org/10.1016/B978-1-78242-458-1.00002-9
160. Hall ME (2020) Coating of technical textiles. In: Horrocks AR, Anand SC (eds) Handbook of technical textiles, 1st ed., vol 17, no 6, ch 8. Texchem Publications, pp 50–54. https://doi.org/10.1533/9781855738966.173
161. Conway R (2016) Coating of textiles. In: Horrocks AR, Anand SC (eds) Handbook of technical textiles: technical textile processes, 2nd ed., vol 1, ch 8. Woodhead Publishing, pp 211–229. https://doi.org/10.1016/B978-1-78242-458-1.00008-X
162. Kopitar D, Marasovic P (2024) Degradation of biodegradable nonwoven mulches in the winter period. Polymers (Basel) 16(16):2279. https://doi.org/10.3390/POLYM16162279
163. Koller M, Heeney D, Mukherjee A (2025) Biodegradability of polyhydroxyalkanoate (PHA) biopolyesters in nature: a review. Biodegradation 36(4):76. https://doi.org/10.1007/S10532-025-10164-Y
164. Kalibová J, Jačka L, Petru J (2016) The effectiveness of jute and coir blankets for erosion control in different field and laboratory conditions. Solid Earth 7(2):469–479. https://doi.org/10.5194/SE-7-469-2016
165. Conway R (2016) Technical textile finishing. In: Horrocks AR, Anand SC (eds) Handbook of technical textiles: technical textile processes, 2nd ed., vol 1, ch 7, Woodhead Publishing, pp 189–210. https://doi.org/10.1016/B978-1-78242-458-1.00007-8

166. Gugumus F (1990) Photo-oxidation and stabilization of polyethylene. In: Scott G (ed) Mechanisms of polymer degradation and stabilisation, 1st ed., ch 6. Springer, Dordrecht, pp 169–209. https://doi.org/10.1007/978-94-011-3838-3_6
167. Brostow W, Lu X, Gencel O, Osmanson AT (2020) Effects of UV stabilizers on polypropylene outdoors. Materials 13(7):1626. https://doi.org/10.3390/MA13071626
168. Gijsman P, Fiorio R (2023) Long term thermo-oxidative degradation and stabilization of polypropylene (PP) and the implications for its recyclability. Polym Degrad Stab 208:110260. https://doi.org/10.1016/J.POLYMDEGRADSTAB.2023.110260
169. Glass RD, Valange BM (1988) Antioxidant 'crossover effect' in oven ageing of polypropylene. Polym Degrad Stab 20(3–4):355–363. https://doi.org/10.1016/0141-3910(88)90078-X
170. Orasugh JT, Temane LT, Kesavan Pillai S, Ray SS (2025) Advancements in antimicrobial textiles: fabrication, mechanisms of action, and applications. ACS Omega 10(13):12772–12816. https://doi.org/10.1021/ACSOMEGA.4C11356
171. Nichterlein M et al (2025) Encapsulated biocides in facade materials impact leaching and UV stability, resulting in lower aquatic toxicity of the eluates. Environ Sci Pollut Res Int 32(27):16324. https://doi.org/10.1007/S11356-025-36647-2
172. Prieto B, Sanmartín P, Silva C, Vázquez-Nion D, Silva B (2014) Deleterious effect plastic-based biocides on back-ventilated granite facades. Int Biodeterior Biodegradation 86:19–24. https://doi.org/10.1016/J.IBIOD.2013.06.018
173. Ghosh J, Rupanty NS, Noor T, Asif TR, Islam T, Reukov V (2025) Functional coatings for textiles: advancements in flame resistance, antimicrobial defense, and self-cleaning performance. RSC Adv 15(14):10984–11022. https://doi.org/10.1039/D5RA01429H
174. Schindler WD, Hauser PJ (2004) Flame-retardant finishes. In: Schindler WD, Hauser PJ (eds) Chemical finishing of textiles, ch 8, pp 98–116. https://doi.org/10.1533/9781845690373.98
175. Hill P, Taylor M, Goswami P, Blackburn R (2016) The route to fluorine-free repellent coatings in outdoor apparel: Consumer use, maintenance and physiological comfort. In: 20th annual green chemistry & engineering conference. White Rose Research Online, Portland, Oregon, USA
176. Sharif R, Mohsin M, Ramzan N, Sardar S, Anam W (2022) Synthesis of bio-based non-fluorinated oil and water repellent finishes for cotton fabric by using palmitic acid, succinic acid, and maleic acid. J Nat Fibers 19(16):14077–14088. https://doi.org/10.1080/15440478.2022.2116141
177. Hadhri M et al (2025) Eco-friendly octylsilane-modified amino-functional silicone coatings for a durable hybrid organic-inorganic water-repellent textile finish. Polymers (Basel) 17(11):1578. https://doi.org/10.3390/POLYM17111578/S1
178. Costa T, Sampaio-Marques B, Neves NM, Aguilar H, Fraga AG (2024) Antimicrobial properties of hindered amine light stabilizers in polymer coating materials and their mechanism of action. Front Bioeng Biotechnol 12:1390513. https://doi.org/10.3389/FBIOE.2024.1390513/FULL
179. Pichon E, De Smet D, Rouster P, Freulings K, Pich A, Bernaerts KV (2023) Bio-based non-isocyanate polyurethane(urea) waterborne dispersions for water resistant textile coatings. Mater Today Chem 34:101822. https://doi.org/10.1016/J.MTCHEM.2023.101822
180. Lacruz A et al (2021) Biobased waterborne polyurethane-ureas modified with poss-oh for fluorine-free hydrophobic textile coatings. Polymers (Basel) 13(20):3526. https://doi.org/10.3390/POLYM13203526/S1
181. Setthayanond J, Suwanruji P, Tooptompong P, Tungtriratanakul S (2022) Colorimetric properties of disperse-dyed polyester fabrics treated with nano TiO_2. Trends Sci 19(14):5067. https://doi.org/10.48048/tis.2022.5067
182. Kinoshita T, Yamamoto T, Luo C, Ide K, Aoki K, Uetsuji Y (2023) Mechanical characterization and design of hybrid reinforced polypropylene composites with talc fillers and cellulose fibers. Compos B Eng 266:110971. https://doi.org/10.1016/J.COMPOSITESB.2023.110971
183. Wang ZY et al (2024) UV protection, antistatic and flame-retardant multifunctional coating of polyester/spandex fabric with carbon black nanoparticles. Polym Degrad Stab 220:110657. https://doi.org/10.1016/J.POLYMDEGRADSTAB.2024.110657

184. Maity S, Singha K, Pandit P (2023) Introduction to functional and technical textiles. In: Maity S, Singha K, Pandit P (eds) Functional and technical textiles, ch 1. Woodhead Publishing, pp 1–30. https://doi.org/10.1016/B978-0-323-91593-9.00021-3
185. Fleury MP et al (2021) Creep behaviour of recycled poly(ethylene) terephthalate non-woven geotextiles. Polymers 13(5):1–17. https://doi.org/10.3390/POLYM13050752
186. Prambauer M, Wendeler C, Weitzenböck J, Burgstaller C (2019) Biodegradable geotextiles—an overview of existing and potential materials. Geotext Geomembr 47(1):48–59. https://doi.org/10.1016/J.GEOTEXMEM.2018.09.006
187. Rollin AL (2004) Long-term performance of geotextiles. In: Demers D, Leahy D, Lefebvre R, Leroueil S, Martel R (eds) 57th Canadian geotechnical conference. BiTech Publishers, Ltd., Quebec, Canada, pp 15–20
188. Brown MH, Badzinski TD, Pardoe E, Ehlebracht M, Maurer-Jones MA (2023) UV light degradation of polylactic acid kickstarts enzymatic hydrolysis. ACS Materials Au 4(1):92. https://doi.org/10.1021/ACSMATERIALSAU.3C00065
189. Yousif E, Haddad R (2013) Photodegradation and photostabilization of polymers, especially polystyrene: review. Springerplus 2(1):398. https://doi.org/10.1186/2193-1801-2-398
190. Karim N et al (2020) Sustainable personal protective clothing for healthcare applications: a review. ACS Nano 14(10):12313. https://doi.org/10.1021/ACSNANO.0C05537
191. Rungruangkitkrai N et al (2024) Water repellent coating in textile, paper and bioplastic polymers: a comprehensive review. Polymers 16(19):2790. https://doi.org/10.3390/POLYM16192790
192. Ferrari M, Cirisano F, Morán MC (2020) Regenerable superhydrophobic coatings for biomedical fabrics. Coatings 10(6):578. https://doi.org/10.3390/COATINGS10060578
193. Vasiliauskienė D, Lukša J, Servienė E, Urbonavičius J (2023) Changes in the bacterial communities of biocomposites with different flame retardants. 13(12):2306. https://doi.org/10.3390/LIFE13122306/S1
194. Xia B, Liu Y, Xing Y, Shi Z, Pan X (2025) Biodegradable medical implants: reshaping future medical practice. Adv Sci (Weinh) 12(35). https://doi.org/10.1002/ADVS.202508014
195. Hegyesi N, Hodosi E, Polyák P, Faludi G, Balogh-Weiser D, Pukánszky B (2020) Controlled degradation of poly-ε-caprolactone for resorbable scaffolds. Colloids Surf B Biointerfaces 186:110678. https://doi.org/10.1016/J.COLSURFB.2019.110678
196. Atalie D, Abtew MA, Zhao L (2025) Recent advances in bio-based nonwoven materials: sustainable production, applications, and circular economy. J Nat Fibers 22(1). https://doi.org/10.1080/15440478.2025.2565661
197. Yao X, Yang X, Lu Y, Qiu Y, Zeng Q (2024) Review of the synthesis and degradation mechanisms of some biodegradable polymers in natural environments. Polymers 17(1):66. https://doi.org/10.3390/POLYM17010066
198. Mondal MIH, Islam MM, Haque MI, Ahmed F (2022) Natural, biodegradable, biocompatible and bioresorbable medical textile materials. In: Medical textiles from natural resources, Elsevier, pp 87–116. https://doi.org/10.1016/B978-0-323-90479-7.00023-3
199. Zaaba NF, Jaafar M (2020) A review on degradation mechanisms of polylactic acid: hydrolytic, photodegradative, microbial, and enzymatic degradation. Wiley. https://doi.org/10.1002/pen.25511
200. Zambrano MC, Pawlak JJ, Daystar J, Ankeny M, Cheng JJ, Venditti RA (2019) Microfibers generated from the laundering of cotton, rayon and polyester based fabrics and their aquatic biodegradation. Mar Pollut Bull 142:394–407. https://doi.org/10.1016/j.marpolbul.2019.02.062
201. Niu J, Zhang X, Pei N, Tian Q (2012) Biodegradability of cellulose fibers and the fabrics in activated sludge. Appl Mech Mater 918–922. https://doi.org/10.4028/www.scientific.net/AMM.217-219.918
202. Seves A, Romano M, Maifreni T, Sora S, Ciferri O (1998) The microbial degradation of silk: a laboratory investigation. Int Biodeterior Biodegrad 42(4):203–211. https://doi.org/10.1016/S0964-8305(98)00050-X

203. Arai T, Freddi G, Innocenti R, Tsukada M (2003) Biodegradation of bombyx mori silk fibroin fibers and films. J Appl Polym Sci 91(4):2383–2390. https://doi.org/10.1002/app.13393
204. Jung JS, Song KH, Kim SH (2021) Biodegradable acetylated kenaf fiber composites. Fibers Polym 22(12):3437–3443. https://doi.org/10.1007/s12221-021-1237-x
205. Sudhakar M, Priyadarshini C, Doble M, Sriyutha Murthy P, Venkatesan R (2007) Marine bacteria mediated degradation of nylon 66 and 6. Int Biodeterior Biodegrad 60(3):144–151. https://doi.org/10.1016/j.ibiod.2007.02.002
206. Zhang J, Wang X, Gong J, Gu Z (2004) A study on the biodegradability of polyethylene terephthalate fiber and diethylene glycol terephthalate. J Appl Polym Sci 93(3):1089–1096. https://doi.org/10.1002/app.20556
207. Li J, Lemstra PJ, Ma P (2022) Chapter 7: can high-performance fibers be(come) bio-based and also biocompostable? Adv Ind Eng Polym Res 5(2):117–132. https://doi.org/10.1016/J.AIEPR.2022.03.002
208. Pfohl P et al (2022) Effect of polymer properties on the biodegradation of polyurethane microplastics. Environ Sci Technol 56(23):16873–16884. https://doi.org/10.1021/ACS.EST.2C05602
209. Giyahchi M, Moghimi H (2023) Aerobic biodegradation of untreated polyester–polyether urethanes by newly isolated yeast strains Exophilia sp. NS-7 and Rhodotorula sp. NS-12. Sci Rep 13(1)5016. https://doi.org/10.1038/s41598-023-31639-z.
210. Reglero Ruiz JA, Trigo-López M, García FC, García JM (2017) Functional aromatic polyamides. Polymers (Basel) 9(9):414. https://doi.org/10.3390/POLYM9090414
211. Ofoegbu SU, Ferreira MGS, Zheludkevich ML (2019) Galvanically stimulated degradation of carbon-fiber reinforced polymer composites: a critical review. Materials 12(4):651. https://doi.org/10.3390/MA12040651
212. Azad HKM, Rahman MZ, Zhu D (2024) Carbon fiber: composites and applications. In: Rangappa SM, Ayyappan V, Manik G, Siengchin S (eds) Synthetic and mineral fibers, their composites and applications, ch 10. Woodhead Publishing, pp 291–319. https://doi.org/10.1016/B978-0-443-13623-8.00010-1.
213. Kukreja K, Panda PK (2024) Aramid fiber, its composites and applications. In: Rangappa SM, Ayyappan V, Manik G, Siengchin S (eds) Synthetic and mineral fibers, their composites and applications, ch 3. Woodhead Publishing, pp 61–99. https://doi.org/10.1016/B978-0-443-13623-8.00003-4
214. Faruk O et al (2023) A comprehensive review of ultrahigh molecular weight polyethylene fibers for applications based on their different preparation techniques. Adv Polym Technol 2023(1):6656692. https://doi.org/10.1155/2023/6656692
215. Rubehn B, Stieglitz T (2010) In vitro evaluation of the long-term stability of polyimide as a material for neural implants. Biomaterials 31(13):3449–3458. https://doi.org/10.1016/j.biomaterials.2010.01.053
216. Shu J, Zhou Z, Liang H, Yang X (2024) Polyimide as a biomedical material: advantages and applications. Nanoscale Adv 6(17):4309–4324. https://doi.org/10.1039/D4NA00292J
217. Zheng XX, Böttger AJ, Jansen KMB, van Turnhout J, van Kranendonk J (2020) Aging of polyphenylene sulfide-glass composite and polysulfone in highly oxidative and strong alkaline environments. Front Mater 7:610440. https://doi.org/10.3389/FMATS.2020.610440
218. Wang Y, Yang T, Fishel K, Benicewicz BC, Chung T-S (2015) Polybenzimidazoles. In: Olabisi O, Adewale KP (eds) Handbook of thermoplastics, ch 19, 2nd ed. CRC Press, Taylor & Francis Group, pp 617–667. https://doi.org/10.13140/RG.2.1.2068.4881
219. Lunt J (1998) Large-scale production, properties and commercial applications of polylactic acid polymers. Polym Degrad Stab 59(1–3):145–152. https://doi.org/10.1016/S0141-3910(97)00148-1
220. Qin Q, Takarada W, Kikutani T (2017) Fiber structure development of phbh through stress-induced crystallization in high-speed melt spinning process. J Fiber Sci Technol 73(2):49–60. https://doi.org/10.2115/fiberst.2017-0007
221. Razzaq S, Shahid S, Nawab Y (2024) Applications and environmental impact of biodegradable polymers in textile industry: a review. Elsevier B.V. https://doi.org/10.1016/j.ijbiomac.2024.136791

222. Zhao X, Cornish K, Vodovotz Y (2020) Narrowing the gap for bioplastic use in food packaging: an update. Environ Sci Technol 54(8):4712–4732. https://doi.org/10.1021/ACS.EST.9B03755
223. Miyaji K, Zhou C, Maeda A, Kobayashi K, Kusumi R (2023) Biodegradation of various forms of cellulose and chitin in natural waters 2 with different salinity. Polym Degrad Stab 215. https://doi.org/10.1016/j.polymdegradstab.2023.110423
224. Chiellini E, Corti A, D'Antone S, Billingham NC (2007) Microbial biomass yield and turnover in soil biodegradation tests: carbon substrate effects. J Polym Environ 15(3):169–178. https://doi.org/10.1007/s10924-007-0057-4
225. Linares AB, Jiménez JC, López P, de Gáscue BR (2019) Biodegradability study by FTIR and DSC of polymers films based on polypropylene and Cassava starch. Orbital 11(2):71–82. https://doi.org/10.17807/orbital.v11i2.1360
226. Holmberg AL, Reno KH, Wool RP, Epps TH (2014) Biobased building blocks for the rational design of renewable block polymers. Soft Matter 10(38):7405–7424. https://doi.org/10.1039/c4sm01220h
227. Warner SB (1995) Fiber science, 1st edn. Prentice Hall, New Jersey
228. Zheng S et al (2021) Implication of surface properties, bacterial motility, and hydrodynamic conditions on bacterial surface sensing and their initial adhesion. Front Bioeng Biotechnol 9:643722. https://doi.org/10.3389/FBIOE.2021.643722
229. Oh JK et al (2018) The influence of surface chemistry on the kinetics and thermodynamics of bacterial adhesion. Sci Rep 8(1):17247. https://doi.org/10.1038/s41598-018-35343-1
230. Chamas A et al (2020) Degradation rates of plastics in the environment. ACS Sustain Chem Eng 8(9):3494–3511. https://doi.org/10.1021/ACSSUSCHEMENG.9B06635
231. Marasović P, Puchalski M, Kopitar D (2025) Effects of field conditions on the degradation of cellulose-based and PLA nonwoven mulches. Sci Rep 15(1):11986. https://doi.org/10.1038/S41598-025-94686-8
232. Hufenus R, Yan Y, Dauner M, Kikutani T (2020) Melt-Spun fibers for textile applications. Materials 13(19):4298. https://doi.org/10.3390/MA13194298
233. Momeni S, Craplewe K, Safder M, Luz S, Sauvageau D, Elias A (2023) Accelerating the biodegradation of poly(lactic acid) through the inclusion of plant fibers: a review of recent advances. ACS Sustain Chem Eng 11(42):15146–15170. https://doi.org/10.1021/ACSSUSCHEMENG.3C04240
234. Takarada W, Sugimoto K, Nakajima H, Visser HA, Gruter GJM, Kikutani T (2021) Melt-spun fibers from bio-based polyester-fiber structure development in high-speed melt spinning of poly(ethylene 2,5-furandicarboxylate) (PEF). Materials 14(5):1172. https://doi.org/10.3390/MA14051172
235. Kopf S, Root A, Heinmaa I, Aristéia de Lima J, Åkesson D, Skrifvars M (2024) Production and characterization of melt-spun poly(3-hydroxybutyrate)/Poly(3-hydroxybutyrate-co-4-hydroxybutyrate) blend monofilaments. ACS Omega 9(25):27415–27427. https://doi.org/10.1021/ACSOMEGA.4C02241
236. Etale A, Onyianta AJ, Turner SR, Eichhorn SJ (2023) Cellulose: a review of water interactions, applications in composites, and water treatment. Chem Rev 123(5):2016–2048. https://doi.org/10.1021/ACS.CHEMREV.2C00477
237. Tan CJ, Lee JJL, Ang BC, Andriyana A, Chagnon G, Sukiman MS (2019) Design of polyurethane fibers: relation between the spinning technique and the resulting fiber topology. J Appl Polym Sci 136(26):47706. https://doi.org/10.1002/APP.47706
238. Sharifzadeh G, Soheilmoghaddam M, Adelnia H, Wahit MU, Arzhandi MRD, Moslehyani A (2020) Biocompatible regenerated cellulose/halloysite nanocomposite fibers. Polym Eng Sci 60(6):1169–1176. https://doi.org/10.1002/PEN.25370
239. Spagnuolo M, Liu L, Coasne B, Obraztsov AN, Zhang Y (2012) Fabrication and degradation of electrospun scaffolds from L-tyrosine-based polyurethane blends for tissue engineering applications. Int Sch Res Notices 2012(1):627420. https://doi.org/10.5402/2012/627420
240. Dias JR, Sousa A, Augusto A, Bártolo PJ, Granja PL (2022) Electrospun polycaprolactone (PCL) degradation: an in vitro and in vivo study. Polym (Basel) 14(16):3397. https://doi.org/10.3390/POLYM14163397

241. Chinnappan BA, Krishnaswamy M, Xu H, Hoque ME (2022) Electrospinning of biomedical nanofibers/nanomembranes: effects of process parameters. Polym (Basel) 14(18):3719. https://doi.org/10.3390/POLYM14183719
242. Calabia BP et al (2013) Biodegradable poly(butylene succinate) composites reinforced by cotton fiber with silane coupling agent. Polym (Basel) 5(1):128–141. https://doi.org/10.3390/polym5010128
243. Liu L, Yu J, Cheng L, Yang X (2009) Biodegradability of poly(butylene succinate) (PBS) composite reinforced with jute fibre. Polym Degrad Stab 94(1):90–94. https://doi.org/10.1016/j.polymdegradstab.2008.10.013
244. Maiti S, Islam MR, Uddin MA, Afroj S, Eichhorn SJ, Karim N (2022) Sustainable fiber-reinforced composites: a review. Wiley. https://doi.org/10.1002/adsu.202200258
245. Mehta S (2018) Optimization of fluorochemical finish concentration for liquid repellency treatment of 100% cotton fabric and resulting physical properties. AATCC J Res 5(5):15–22. https://doi.org/10.14504/AJR.5.5.3
246. Mehta S (2020) An experimental investigation on optimizing liquid repellency of fluorochemical urethane finish and its effect on the physical properties of polyester/cotton blended fabric. Fibers 8(12):1–14. https://doi.org/10.3390/fib8120072
247. Schindler WD, Hauser PJ (2004) Repellent finishes. In: Schindler WD, Hauser PJ (eds) Chemical finishing of textiles, vol 2, ch 6. Woodhead Publishing, pp 74–86. https://doi.org/10.1533/9781845690373.74
248. Fitzgerald JF, Kumar AS (2014) Biologic versus synthetic mesh reinforcement: what are the pros and cons? Clin Colon Rectal Surg 27(4):140. https://doi.org/10.1055/S-0034-1394155
249. Baylón K, Rodríguez-Camarillo P, Elías-Zúñiga A, Díaz-Elizondo JA, Gilkerson R, Lozano K (2017) Past, present and future of surgical meshes: a review. Membr (Basel) 7(3):47. https://doi.org/10.3390/MEMBRANES7030047
250. Najm A, Niculescu AG, Gaspar BS, Grumezescu AM, Beuran M (2023) A review of abdominal meshes for hernia repair-current status and emerging solutions. Materials (Basel) 16(22). https://doi.org/10.3390/MA16227124
251. Krasselt K, Frommelt C, Brunner R, Rauscher FG, Francke M, Körber N (2020) Various cross-linking methods inhibit the collagenase I degradation of rabbit scleral tissue. BMC Ophthalmol 20(1):488. https://doi.org/10.1186/S12886-020-01751-Z
252. Lazić V et al (2015) Negative influence of Ag and TiO_2 nanoparticles on biodegradation of cotton fabrics. Cellulose 22(2):1365–1378. https://doi.org/10.1007/s10570-015-0549-7
253. Klemenčič D, Simončič B, Tomšič B, Orel B (2010) Biodegradation of silver functionalised cellulose fibres. Carbohydr Polym 80(2):426–435. https://doi.org/10.1016/j.carbpol.2009.11.049
254. Akhir MAM, Ramakrishnan S, Mariatti M (2025) Anti-hydrolysis, UV absorber, and photostabilizer additives effect on degradation of poly(butylene adipate-co-terephthalate) biocomposite film for mulching application. Polym Adv Technol 36(7):e70266. https://doi.org/10.1002/PAT.70266
255. Rizzarelli P, Rapisarda M, Ascione L, Innocenti FD, Degli Innocenti F (2021) Influence of photo-oxidation on the performance and soil degradation of oxo- and biodegradable polymer-based items for agricultural applications. Polym Degrad Stab 188:109578. https://doi.org/10.1016/J.POLYMDEGRADSTAB.2021.109578
256. Chang BP, Mohanty AK, Misra M (2020) Studies on durability of sustainable biobased composites: a review. RSC Adv 10(31):17955–17999. https://doi.org/10.1039/C9RA09554C
257. Li L, Frey M, Browning KJ (2010) Biodegradability study on cotton and polyester fabrics. J Eng Fiber Fabr 5(4). https://doi.org/10.1177/155892501000500406.
258. Kim MC, Masuoka T (2009) Degradation properties of PLA and PHBV films treated with CO_2-plasma. React Funct Polym 69(5):287–292. https://doi.org/10.1016/j.reactfunctpolym.2009.01.013
259. Tang KHD (2023) State of the art in textile waste management: a review. Textiles 3(4):454–467. https://doi.org/10.3390/TEXTILES3040027

260. Amicarelli V, Bux C (2022) Quantifying textile streams and recycling prospects in Europe by material flow analysis. Environ Impact Assess Rev 97:106878. https://doi.org/10.1016/J.EIAR.2022.106878
261. Sandin G, Peters GM (2018) Environmental impact of textile reuse and recycling—a review. J Clean Prod 184:353–365. https://doi.org/10.1016/J.JCLEPRO.2018.02.266
262. Biyada S et al (2021) Microbial community dynamics in the mesophilic and thermophilic phases of textile waste composting identified through next-generation sequencing. Sci Rep 11(1):23624. https://doi.org/10.1038/s41598-021-03191-1
263. Mouhoubi R, Lasschuijt M, Ramon Carrasco S, Gojzewski H, Wurm FR (2022) End-of-life biodegradation? How to assess the composting of polyesters in the lab and the field. Waste Manag 154:36–48. https://doi.org/10.1016/J.WASMAN.2022.09.025
264. Wojnowska-Baryła I, Bernat K, Zaborowska M (2022) Strategies of recovery and organic recycling used in textile waste management. Int J Environ Res Public Health 19(10):5859. https://doi.org/10.3390/IJERPH19105859
265. Barletta M et al (2022) Poly(butylene succinate) (PBS): materials, processing, and industrial applications. Prog Polym Sci 132:101579. https://doi.org/10.1016/J.PROGPOLYMSCI.2022.101579
266. Kopitar D, Marasovic P, Jugov N, Schwarz I (2022) Biodegradable nonwoven agrotextile and films—a review. Polymers 14(11):2272. https://doi.org/10.3390/POLYM14112272
267. Mörtl M et al (2024) Biodegradation assessment of bioplastic carrier bags under industrial-scale composting conditions. Polymers 16(24):3450. https://doi.org/10.3390/POLYM16243450
268. Borelbach P, Kopitzky R, Dahringer J, Gutmann P (2023) Degradation behavior of biodegradable man-made fibers in natural soil and in compost. Polymers (Basel) 15(13):2959. https://doi.org/10.3390/POLYM15132959/S1
269. Nayak R, Singh A, Thai VV, Grover T, Tran DTM (2025) Recent sustainability trends in composting textiles: process, factors, benefits and challenges. Results Eng 28:107694. https://doi.org/10.1016/J.RINENG.2025.107694
270. Meereboer KW, Misra M, Mohanty AK (2020) Review of recent advances in the biodegradability of polyhydroxyalkanoate (PHA) bioplastics and their composites. Green Chem 22(17):5519–5558. https://doi.org/10.1039/D0GC01647K
271. Trofimchuk E et al (2023) Degradation of structurally modified polylactide under the controlled composting of food waste. Polymers 2023 15(19):4017. https://doi.org/10.3390/POLYM15194017
272. Schick S, Groten R, Seide GH (2023) Performance spectrum of home-compostable biopolymer fibers compared to a petrochemical alternative. Polymers (Basel) 15(6):1372. https://doi.org/10.3390/POLYM15061372
273. Gadaleta G, Andrade-Chapal JC, López-Ibáñez S, Mozo-Toledo M, Navarro-Calderón Á (2025) Biodegradability of bioplastics in managed and unmanaged environments: a comprehensive review. Materials 18(10):2382. https://doi.org/10.3390/MA18102382
274. Fogašová M et al (2022) PLA/PHB-based materials fully biodegradable under both industrial and home-composting conditions. Polymers (Basel) 14(19):4113. https://doi.org/10.3390/POLYM14194113
275. Patti A, Cicala G, Acierno D (2020) Eco-sustainability of the textile production: waste recovery and current recycling in the composites world. Polymers 13(1):134. https://doi.org/10.3390/POLYM13010134
276. Pischedda A, Tosin M, Degli-Innocenti F (2019) Biodegradation of plastics in soil: the effect of temperature. Polym Degrad Stab 170:109017. https://doi.org/10.1016/J.POLYMDEGRADSTAB.2019.109017
277. Royer SJ, Greco F, Kogler M, Deheyn DD (2023) Not so biodegradable: polylactic acid and cellulose/plastic blend textiles lack fast biodegradation in marine waters. PLoS ONE 18(5):e0284681. https://doi.org/10.1371/JOURNAL.PONE.0284681
278. Hubbe MA et al (2025) Biodegradability of cellulose fibers, films, and particles: a review. BioResources 20(1):2391–2458. https://doi.org/10.15376/biores.20.1.Hubbe

279. Brunšek R, Kopitar D, Schwarz I, Marasović P (2023) Biodegradation properties of cellulose fibers and PLA biopolymer. Polymers 15(17):3532. https://doi.org/10.3390/POLYM15173532
280. Poli V, Lavagnolo MC, Basaglia M, Bonato T, Zanatta S, Modesti M (2025) Assessment of the biodegradability of polylactic acid (PLA) in freshwater using EN ISO 14851:2019: challenges and outcomes. J Hazard Mater 491:137974. https://doi.org/10.1016/J.JHAZMAT.2025.137974
281. Chamley A et al (2024) Polymer material biodegradation in the deep sea. A review. Sci Total Environ 957:177637. https://doi.org/10.1016/J.SCITOTENV.2024.177637
282. Lv S, Li Y, Zhao S, Shao Z (2024) Biodegradation of typical plastics: from microbial diversity to metabolic mechanisms. Int J Molecular Sci 25(1):593. https://doi.org/10.3390/IJMS25010593
283. Eronen-Rasimus EL, Näkki PP, Kaartokallio HP (2022) Degradation rates and bacterial community compositions vary among commonly used bioplastic materials in a Brackish marine environment. Environ Sci Technol 56(22):15760–15769. https://doi.org/10.1021/acs.est.2c06280
284. Dilkes-Hoffman LS, Lant PA, Laycock B, Pratt S (2019) The rate of biodegradation of PHA bioplastics in the marine environment: a meta-study. Mar Pollut Bull 142:15–24. https://doi.org/10.1016/J.MARPOLBUL.2019.03.020
285. Magalhães S et al (2024) Insights on microplastic contamination from municipal and textile industry effluents and their removal using a cellulose-based approach. Polymers (Basel) 16(19):2803. https://doi.org/10.3390/POLYM16192803/S1
286. Krzan A, Hemjinda S, Miertus S, Corti A, Chiellini E (2006) Standardization and certification in the area of environmentally degradable plastics. Polym Degrad Stab 91(12):2819–2833. https://doi.org/10.1016/J.POLYMDEGRADSTAB.2006.04.034
287. Yagi H, Ninomiya F, Funabashi M, Kunioka M (2009) Anaerobic biodegradation tests of poly(lactic acid) under mesophilic and thermophilic conditions using a new evaluation system for methane fermentation in anaerobic sludge. Int J Molecular Sci 10(9):3824–3835. https://doi.org/10.3390/IJMS10093824
288. Kolstad JJ, Vink ETH, De Wilde B, Debeer L (2012) Assessment of anaerobic degradation of Ingeo™ polylactides under accelerated landfill conditions. Polym Degrad Stab 97(7):1131–1141. https://doi.org/10.1016/J.POLYMDEGRADSTAB.2012.04.003
289. Muller R-J (2005) Biodegradability of polymers: regulations and methods for testing. In: Steinbüchel A (ed) Biopolymers online. Wiley. https://doi.org/10.1002/3527600035.bpola012
290. Lin C, Cheruiyot NK, Bui XT, Ngo HH (2022) Composting and its application in bioremediation of organic contaminants. Taylor and Francis Ltd. https://doi.org/10.1080/21655979.2021.2017624
291. Salazar VLP, Leão AL, Rosa DS, Gomez JGC, Alli RCP (2011) Biodegradation of coir and sisal applied in the automotive industry. J Polym Environ 19(3):677–688. https://doi.org/10.1007/s10924-011-0315-3
292. Han Y, Wang R, Wang D, Luan Y (2024) Enzymatic degradation of synthetic plastics by hydrolases/oxidoreductases. Int Biodeterior Biodegradation 189:105746. https://doi.org/10.1016/J.IBIOD.2024.105746
293. Delacuvellerie A et al (2021) Microbial biofilm composition and polymer degradation of compostable and non-compostable plastics immersed in the marine environment. J Hazard Mater 419. https://doi.org/10.1016/j.jhazmat.2021.126526
294. Jamee R, Siddique R (2019) Biodegradation of synthetic dyes of textile effluent by microorganisms: an environmentally and economically sustainable approach. Eur J Microbiol Immunol (Bp) 9(4):114. https://doi.org/10.1556/1886.2019.00018
295. Cavaco-Paulo A (1998) Mechanism of cellulase action in textile processes. Carbohydr Polym 37(3):273–277. https://doi.org/10.1016/S0144-8617(98)00070-8
296. Shalem A, Yehezkeli O, Fishman A (2024) Enzymatic degradation of polylactic acid (PLA). Appl Microbiol Biotechnol 108(1):413. https://doi.org/10.1007/S00253-024-13212-4

297. Temporiti MEE, Nicola L, Nielsen E, Tosi S (2022) Fungal enzymes involved in plastics biodegradation. Microorganisms 10(6):1180. https://doi.org/10.3390/MICROORGANISMS10061180
298. Folino A, Pangallo D, Calabrò PS (2023) Assessing bioplastics biodegradability by standard and research methods: current trends and open issues. J Environ Chem Eng 11(2):109424. https://doi.org/10.1016/J.JECE.2023.109424
299. Bucişcanu I-I, Pruneanu M (2022) Key elements of sustainability in the field of technical textiles. In: International symposium "technical textiles—present and future" Sciendo, pp 122–129. https://doi.org/10.2478/9788366675735-021
300. Adeleh S, Bol R, Becker T, Herres-Pawlis S, Vereecken H, Pütz T (2024) Radiolabeling for polymers degradation studies: opportunities and challenges ahead. Polym Degrad Stab 230:111053. https://doi.org/10.1016/J.POLYMDEGRADSTAB.2024.111053
301. Teggers EM et al (2025) Diversifying endpoints in biodegradation testing of microplastics. Environ Sci Europe 37(1):65. https://doi.org/10.1186/S12302-025-01096-8
302. Julinová M et al (2023) Comprehensive biodegradation analysis of chemically modified poly(3-hydroxybutyrate) materials with different crystal structures. Biomacromol 24(11):4939–4957. https://doi.org/10.1021/ACS.BIOMAC.3C00623
303. Clarkson MO et al (2024) A review of measurement for quantification of carbon dioxide removal by enhanced weathering in soil. Front Clim 6:1345224. https://doi.org/10.3389/FCLIM.2024.1345224
304. Mndlovu H, Kumar P, du Toit LC, Choonara YE (2024) A review of biomaterial degradation assessment approaches employed in the biomedical field. NPJ Mater Degrad 8(1):66. https://doi.org/10.1038/s41529-024-00487-1
305. Siakeng R, Jawaid M, Asim M, Siengchin S (2020) Accelerated weathering and soil burial effect on biodegradability, colour and texture of coir/pineapple leaf fibres/PLA biocomposites. Polymers 12(2):458. https://doi.org/10.3390/POLYM12020458
306. ISO (2018) ISO 14855-2:2018: determination of the ultimate aerobic biodegradability of plastic materials under controlled composting conditions—method by analysis of evolved carbon dioxide—part 2: gravimetric measurement of carbon dioxide evolved in a laboratory-scale test. https://www.iso.org/home.html
307. ASTM International (2021) ASTM D5338-15(2021), standard test method for determining aerobic biodegradation of plastic materials under controlled composting conditions, incorporating thermophilic temperatures. https://doi.org/10.1520/D5338-15R21.
308. ISO (2023) ISO 20200:2023—plastics—determination of the degree of disintegration of plastic materials under composting conditions in a laboratory-scale test. https://www.iso.org/home.html
309. ISO (2021) ISO 16929:2021—plastics—determination of the degree of disintegration of plastic materials under defined composting conditions in a pilot-scale test. https://www.iso.org/home.html
310. ISO (2025a) ISO 21701:2025—textiles—test method for accelerated hydrolysis of textile materials and biodegradation under controlled composting conditions of the resulting hydrolysate. https://www.iso.org/home.html
311. ISO (2025b) ISO/CD 17952—textiles—test method for determination of physical degradation rate of textile materials under simulated composting conditions in a laboratory—scale test. https://www.iso.org/home.html
312. AS (2010) AS 5810-2010—biodegradable plastics—biodegradable plastics suitable for home composting. https://www.standards.org.au/
313. AFNOR (2015) Standard NF T51-800, specifications for plastics suitable for home composting. https://www.boutique.afnor.org/en-gb/
314. ISO (2019) ISO 17556:2019—plastics—determination of the ultimate aerobic biodegradability of plastic materials in soil by measuring the oxygen demand in a respirometer or the amount of carbon dioxide evolved. https://www.iso.org/home.html
315. BSI (2018) BS EN 17033:2018 plastics. Biodegradable mulch films for use in agriculture and horticulture. Requirements and test methods. https://www.bsigroup.com/

316. ASTM International (2018) ASTM D5988-18, standard test method for determining aerobic biodegradation of plastic materials in soil. https://doi.org/10.1520/D5988-18.
317. ISO (2001) ISO 11721-1:2001—textiles—determination of resistance of cellulose-containing textiles to micro-organisms—soil burial test—part 1: assessment of rot-retardant finishing. https://www.iso.org/home.html
318. ISO (2003) ISO 11721-2:2003—textiles—determination of the resistance of cellulose-containing textiles to micro-organisms—soil burial test—part 2: identification of long-term resistance of a rot retardant finish. https://www.iso.org/home.html
319. AATCC (2017) AATCC TM30-2017, test method for antifungal activity, assessment on textile materials: mildew and rot resistance of textile materials. https://www.aatcc.org/
320. ISO (2019) ISO 14851:2019—determination of the ultimate aerobic biodegradability of plastic materials in an aqueous medium—method by measuring the oxygen demand in a closed respirometer. https://www.iso.org/home.html
321. ISO (2021) ISO 14852:2021—determination of the ultimate aerobic biodegradability of plastic materials in an aqueous medium—method by analysis of evolved carbon dioxide. https://www.iso.org/home.html
322. BSI (2006) BS EN 14987:2006 plastics. Evaluation of disposability in waste water treatment plants. Test scheme for final acceptance and specifications. https://www.bsigroup.com/
323. DIN (2023) DIN SPEC 4872:2023-02, test method for textiles—determination of fibre release during washing and aerobic degradation level in aqueous medium in consideration of ecotoxicity. DIN Media GmbH, Berlin. https://doi.org/10.31030/3404156.
324. ASTM International (2017) ASTM D6691-17, standard test method for determining aerobic biodegradation of plastic materials in the marine environment by a defined microbial consortium or natural sea water inoculum. https://doi.org/10.1520/D6691-17
325. ISO (2020) ISO 22403:2020—plastics—assessment of the intrinsic biodegradability of materials exposed to marine inocula under mesophilic aerobic laboratory conditions—test methods and requirements. https://www.iso.org/home.html
326. ISO (2020) ISO 19679:2020—plastics—determination of aerobic biodegradation of non-floating plastic materials in a seawater/sediment interface—method by analysis of evolved carbon dioxide. https://www.iso.org/home.html
327. ISO (2016) ISO 18830:2016—plastics—determination of aerobic biodegradation of non-floating plastic materials in a seawater/sandy sediment interface—method by measuring the oxygen demand in closed respirometer. https://www.iso.org/home.html
328. ISO (2019) ISO 22404:2019—plastics—determination of the aerobic biodegradation of non-floating materials exposed to marine sediment—method by analysis of evolved carbon dioxide. https://www.iso.org/home.html
329. ISO (2014) ISO 15985:2014—plastics—determination of the ultimate anaerobic biodegradation under high-solids anaerobic-digestion conditions—method by analysis of released biogas. https://www.iso.org/home.html
330. ASTM International (2018) ASTM D5511-18, standard test method for determining anaerobic biodegradation of plastic materials under high-solids anaerobic-digestion conditions. https://doi.org/10.1520/D5511-18
331. ASTM International (2018) ASTM D5526-18, standard test method for determining anaerobic biodegradation of plastic materials under accelerated landfill conditions (withdrawn 2025). https://www.astm.org/
332. ISO (2016) ISO 14853:2016—plastics—determination of the ultimate anaerobic biodegradation of plastic materials in an aqueous system—method by measurement of biogas production. https://www.iso.org/home.html
333. ISO (2019) ISO 13975:2019—plastics—determination of the ultimate anaerobic biodegradation of plastic materials in controlled slurry digestion systems—method by measurement of biogas production. https://www.iso.org/home.html
334. Shah AA, Hasan F, Hameed A, Ahmed S (2008) Biological degradation of plastics: a comprehensive review. Biotechnol Adv 26(3):246–265. https://doi.org/10.1016/J.BIOTECHADV.2007.12.005

335. Andrady AL et al (2022) Oxidation and fragmentation of plastics in a changing environment; from UV-radiation to biological degradation. Sci Total Environ 851(Pt 2):158022. https://doi.org/10.1016/J.SCITOTENV.2022.158022
336. ASTM International (2021) ASTM D4329-21, standard practice for fluorescent ultraviolet (UV) lamp apparatus exposure of plastics. https://doi.org/10.1520/D4329-21
337. ASTM International (2023) ASTM G154-23, standard practice for operating fluorescent ultraviolet (UV) lamp apparatus for exposure of materials. https://doi.org/10.1520/G0154-23
338. AATCC (2023) AATCC TM186 test method for weather resistance: UV light and moisture exposure. https://www.aatcc.org/
339. AATCC (2023) AATCC TM111 test method for weather resistance of textiles. https://www.aatcc.org/
340. ISO (2025) ISO 877:2025—plastics—methods of exposure to solar radiation—part 1/2/3. https://www.iso.org/home.html
341. ISO (2024) ISO 4892-3:2024—plastics—methods of exposure to laboratory light sources—part 3: fluorescent UV lamps. https://www.iso.org/home.html
342. BSI (2003) BS EN 14119:2003 testing of textiles. Evaluation of the action of microfungi. https://www.bsigroup.com/
343. Min K, Cuiffi JD, Mathers RT (2020) Ranking environmental degradation trends of plastic marine debris based on physical properties and molecular structure. Nat Commun 11(1):727. https://doi.org/10.1038/s41467-020-14538-z
344. Takekoshi S, Takano K, Matoba Y, Mukumoto M, Tachibana A (2022) Establishing a ready biodegradability test system using OxiTop® to evaluate chemical fate in a realistic environment. J Pestic Sci 47(1):35. https://doi.org/10.1584/JPESTICS.D21-046
345. Davenport R et al (2022) Scientific concepts and methods for moving persistence assessments into the 21st century. Integr Environ Assess Manag 18(6):1454–1487. https://doi.org/10.1002/ieam.4575
346. Beiras R, López-Ibáñez S (2023) A practical tool for the assessment of polymer biodegradability in marine environments guides the development of truly biodegradable plastics. Polymers 15(4):974. https://doi.org/10.3390/POLYM15040974
347. Nam S et al (2024) Soil biodegradation resistance of cotton fiber doped with interior and exterior silver nanoparticles. ACS Omega. https://doi.org/10.1021/ACSOMEGA.3C09390
348. Totito TC, Laatikainen K, Bode-Aluko C, Pereao O, Petrik L (2023) Fabrication and characterization of electrospun waste polyethylene terephthalate blended with chitosan: a potential single-use material. Polymers 15(2):442. https://doi.org/10.3390/POLYM15020442
349. Karim FE, Islam MR, Ahmed R, Siddique AB, Begum HA (2023) Extraction and characterization of a newly developed cellulose enriched sustainable natural fiber from the epidermis of Mikania micrantha. Heliyon 9(9):e19360. https://doi.org/10.1016/J.HELIYON.2023.E19360
350. Khurshid MF, Hasan MMB, Abdkader A, Cherif C (2022) Processing of waste carbon and polyamide fibers for high performance thermoplastic composites: a novel manufacturing technology for unidirectional tapes structure. J Ind Text 51(4):7256S-7276S. https://doi.org/10.1177/15280837221077705
351. Wu HC, Wu SR, Yang TCK, Yang JC (2018) A facile measurement for monitoring dragline silk dope concentration in nephila pilipes upon spinning. Materials 11(10):1951. https://doi.org/10.3390/MA11101951
352. Briassoulis D, Dejean C (2010) Critical review of norms and standards for biodegradable agricultural plastics part I. Biodegradation in soil. J Polym Environ 18(3):384–400. https://doi.org/10.1007/S10924-010-0168-1
353. Kim YH et al (2021) Biodegradable hot melt adhesive based on partially saponified polyvinyl acetate/cellulose diacetate blend. J Appl Polym Sci 138(35):50867. https://doi.org/10.1002/APP.50867
354. Narancic T et al (2018) Biodegradable plastic blends create new possibilities for end-of-life management of plastics but they are not a panacea for plastic pollution. Environ Sci Technol 52(18):10441–10452. https://doi.org/10.1021/ACS.EST.8B02963

355. Le Thuy T, Hoang TD, Hoang VH, Nguyen MK (2025) A Review on flame retardants in soils: occurrence, environmental impact, health risks, remediation strategies, and future perspectives. Toxics 13(3):228. https://doi.org/10.3390/TOXICS13030228
356. Holanda FSR et al (2023) Comparison of mechanical performance of different geotextiles manufactured from natural fibers subjected to environmental biodegradation. Int J Adv Manufact Technol 129(9):3863–3873. https://doi.org/10.1007/S00170-023-12549-W
357. Luo J et al (2022) Elastic vitrimers: Beyond thermoplastic and thermoset elastomers. Matter 5(5):1391–1422. https://doi.org/10.1016/J.MATT.2022.04.007
358. Rathinamoorthy R, Raja-Balasaraswathi S, Madhubashini S, Prakalya A, Rakshana JB, Shathvika S (2023) Investigation on microfiber release from elastane blended fabrics and its environmental significance. Sci Total Environ 903:166553. https://doi.org/10.1016/J.SCITOTENV.2023.166553
359. Giridev VR, Sudha S (2024) Scope of sustainable materials in medical textiles. In: Sundaramoorthy S, Kumar SKS, Chavhan MV (eds) Textile materials for good health and well-being, ch 19. Springer, Singapore, pp 337–348. https://doi.org/10.1007/978-981-97-6496-9_19
360. Gorgieva S, Zemljić LF, Strnad S, Kokol V (2018) Textile-based biomaterials for surgical applications. Fund Biomater Polym 14:179–215. https://doi.org/10.1016/B978-0-08-102194-1.00009-8
361. Heidari BS et al (2024) Emerging materials and technologies for advancing bioresorbable surgical meshes. Acta Biomater 184:1–21. https://doi.org/10.1016/J.ACTBIO.2024.06.012
362. Zamparas M, Tzivras D, Dracopoulos V, Ioannides T (2020) Application of sorbents for oil spill cleanup focusing on natural-based modified materials: a review. Molecules 25(19):4522. https://doi.org/10.3390/MOLECULES25194522
363. Zawierucha I et al (2022) Ecotoxicity and bioremediation potential assessment of soil from oil refinery station area. J Environ Health Sci Eng 20(1):337. https://doi.org/10.1007/S40201-021-00780-0
364. Liu P et al (2024) Mechanically triggered on-demand degradation of polymers synthesized by radical polymerizations. Nat Chem 16(7):1184–1192. https://doi.org/10.1038/s41557-024-01508-x
365. Al-Madhagi H (2025) Microplastics toxicology and bioremediation strategies for a sustainable future: a comprehensive review. Green Chem Lett Rev 18(1):390. https://doi.org/10.1080/17518253.2025.2535390

Ultraviolet Protective Textiles: Mechanisms, Material Parameters and Sustainable Approaches

Sanjida Sultana, Tanzima Zoha Chowdhury, Sushama Saha Swati, Zinia Yasmin, and Mahfuza Pervin

Abstract Ultraviolet (UV) radiation provides a substantial risk to human health, causing eye damage, skin cancer and premature ageing. These growing concerns have led to increasing demand for fabrics with efficient UV protection. The first section of this chapter discusses the impact of UV exposure on health and the need for protective textile solutions. These methods to improve textile resistance against UV light include UV-absorbing polymers, chemical treatments, and special surface coatings. The chapter presents information available on the interaction of UV radiation with textiles either through absorption, reflection, or scattering and how this is related to fiber type, fabric structures, colorations, and finishing techniques for levels of protection. It also addresses the longevity of UV-protective properties, especially how well they stand up over time to frequent washings, wear and exposure to elements. Various testing procedures and standards, such as those for the Ultraviolet Protection Factor (UPF), are evaluated to determine their durability and effectiveness of these protective characteristics. Increasing focus is being placed on eco-friendly UV-protective textile technologies. This chapter emphasizes the use of bio-based UV absorbers, natural dyes, eco-friendly nanoparticles, and green finishing methods that lower chemical toxicity and environmental impact. It also discusses the importance of durability, reduced chemical leaching, recyclability, and longer product lifespan

The original version of the chapter has been revised. A correction to this chapter can be found at https://doi.org/10.1007/978-3-032-29974-1_7

S. Sultana · T. Z. Chowdhury (✉) · S. S. Swati · Z. Yasmin · M. Pervin
Department of Textile Engineering, Primeasia University, Dhaka, Bangladesh
e-mail: tanzima.zoha@primeasia.edu.bd

S. Sultana
e-mail: sanjida.sultana@primeasia.edu.bd

S. S. Swati
e-mail: sushama.swati@primeasia.edu.bd

Z. Yasmin
e-mail: zinia.yasmin@primeasia.edu.bd

M. Pervin
e-mail: mahfuza.pervin@primeasia.edu.bd

S. S. Muthu (ed.), *Technical Textiles for Sustainability*, Sustainable Textiles: Production, Processing, Manufacturing & Chemistry, https://doi.org/10.1007/978-3-032-29974-1_5

as essential factors for sustainable UV-protective textiles. These methods help reduce resource use and textile waste while ensuring strong protective performance.

Keywords Ultraviolet protection factor · Sustainability · Radiation · Transmission · Physical blocking · Chemical absorption · Phyto-chemicals · Porosity

1 Introduction

Ultraviolet (UV) radiation has been an ever-increasing threat owing to its accelerating adverse impact on the health of people and organic materials [1]. Even though UV light covers wavelengths from 40–400 nm, only UVA (320–400 nm) and UVB (290–320 nm) can reach the surface of the Earth due to the filtering action of the stratospheric ozone layer, which restricts shorter, more harmful wavelengths. Even so, these residual UV rays can still cause serious biological harm, to the point that they contribute to skin cancer, immunosuppression, cataracts, early skin aging, as well as the onset of Alzheimer's disease and inflammatory disorders. Additionally, UV exposure accelerates and speeds up the discoloration and mechanical weakening of textiles, plastics, paints, and wood. This issue is made worse by the continuous ozone depletion brought on by man-made pollutants like nitric oxide, nitrous oxide, and organohalogen compounds.

So, the development of efficient UV-blocking materials is key to public health and environmental protection. Due to their soft durability, flexibility and comfort, textiles have been used for sun protective barriers since long ago, and the World Health Organization (WHO) recommends full coverage clothing as a UV-protective measure in outdoor environments [2]. But untreated textiles—particularly lightweight summer materials—typically don't provide adequate UV protection. Additionally, these fabrics need to not only protect the skin but also their own fibers and dyes from UV deterioration in order to maintain performance during demanding applications, such as tents, awnings and parachutes [3]. These limitations highlight the need for advanced UV-blocking treatments. Recent research has therefore focused on enhancing textile performance through nanostructured surface coatings, assessed using the Ultraviolet Protection Factor (UPF), a metric analogous to the SPF used for sunscreens (AATCC 183-2004; AS/NZS 4399:1996; EN 13,758–1:2002). Emerging evidence also underscores the significance of UVA radiation in photo-aging and its role in amplifying UVB-induced carcinogenesis, reinforcing the importance of broad-spectrum UV-protective textile technologies [4].

Ultraviolet (UV), visible, and infrared light are all considered optical radiation. UV radiation is separated into UV-A, UV-B, and UV-C, and its wavelengths range from 100 to 400 nm. Although artificial sources like tanning beds and office lamps are becoming more significant, the sun is the primary source of UV exposure. The atmosphere virtually completely blocks UV-C from the sun. In addition to causing sunburn, skin cancer, and immunosuppression, UV-B is essential for the synthesis of vitamin D, which may reduce the risk of several malignancies.

2 Effects of UV on the Human Body

Ultraviolet (UV), visible, and infrared light are all considered optical radiation. UV radiation is separated into UV-A, UV-B, and UV-C, and its wavelengths range from 100 to 400 nm [5] (Table 1).Although artificial sources like tanning beds and office lamps are becoming more significant nowadays, the sun is the primary source of UV exposure. The atmosphere virtually completely blocks UV-C from the sun. In addition to causing sunburn, skin cancer, and immunosuppression, UV-B is essential for the synthesis of vitamin D, which may reduce the risk of several malignancies [6]. But ultraviolet radiation (UVR) may cause three primary types of skin cancer: malignant melanoma of the skin, squalor cell carcinoma, and basal cell carcinoma. Melanoma rates are still rising, general sun protection compliance is uneven, and public understanding of the risk is inadequate [7].

Solar radiation exposure can have a significant impact on human health. The skin is mainly the medium for health hazards. Especially the effect on human skin is crucial in this situation. The optical characteristics that influence the nature of these interactions are absorption, scattering, and reflection. The skin is an inhomogeneous, multilayered organ. The biological properties of its primary constituents and how they impact light absorption and propagation are depicted in Fig. 1 [9].

Sunburn inflammation (erythema), tanning, and local or systemic immunosuppression are the main acute consequences of UV radiation on healthy human skin. UV exposure damages DNA at the molecular level, resulting in cyclobutane pyrimidine dimers and (6–4) photoproducts, which are often fixed by nucleotide excision

Table 1 Effects of ultraviolet radiation on the human body [8]

Sources of UVR	Wave length (λ) nm	Relative intensity (%)	Avg photon energy (KJ/mol)	(%) of UV reaches on earth surface	UVR-induced adverse effects on human health
UV-C	280_100	0.5	400	About 0% (little)	Extremely dangerous and impenetrable by the ozone layer
UV-B	315_280	2.4	350	1%_10%	Skin cancer, cataracts, sunburn, painful erythema, and blistering
UV-A	400_315	3.2	315	95%	Premature aging, wrinkling of the skin, implicated in skin cancer, sun tanning, scaling, and dryness

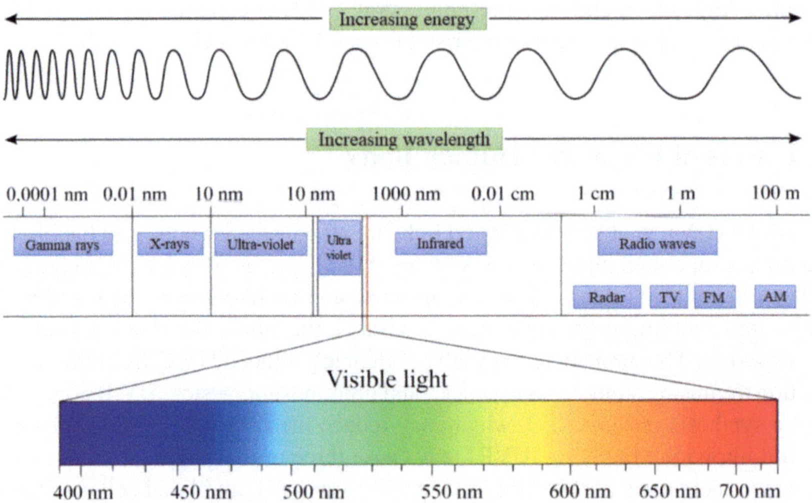

Fig. 1 Bands of the electromagnetic spectrum by common names [6]

repair (NER). Prolonged UV exposure leads to immunosuppression, photoaging, and ultimately, photo-carcinogenesis [10].

2.1 The Use of Textile Fabrics for UV Protection

Clothing designed for UV protection should be worn in all scenarios where exposure to any type of UV radiation is anticipated: adolescents, individuals working outdoors, athletes, people working with advertising lights, curing plastics, sterilization equipment, medical professionals, currency verification lights, and industrial welding [11].

Clothing can effectively shield against harmful UV radiation and plays a crucial role in preventing various dangerous effects associated with these rays. Dermatologists have established a direct link between the number of tumors on a patient's skin and their choice of clothing. The ability of fabrics to protect against UV radiation depends on the amount of UVR they allow to pass through. When UV radiation falls on a fabric, some of it gets transmitted without obstruction, some is scattered, and some is absorbed by the material and transformed into other types of energy [Fig. 2].

Generally, a textile is deemed to provide acceptable UV protection if it transmits less than 6% of total incident UVR, while a transmission rate below 2.5% is regarded as excellent [12].

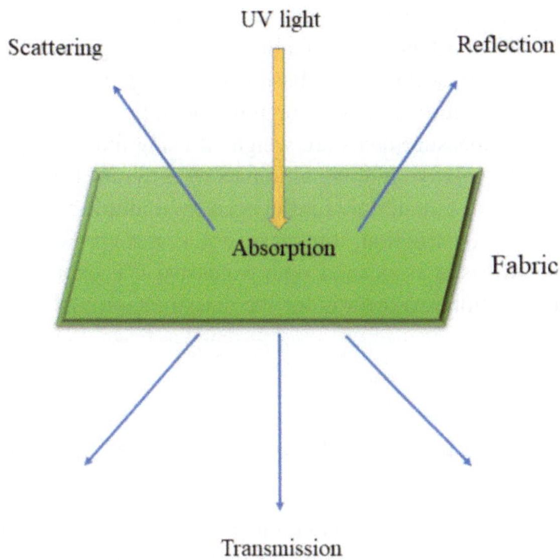

Fig. 2 UV transmittance characteristics of a textile material

2.2 *Different Parameters for UV-Protective Textile Fabrics*

2.2.1 UV Index (UVI)

The ultraviolet (UV) index (UVI) is a standardized metric that measures the intensity of solar ultraviolet radiation that reaches the Earth's surface and its possible biological impacts. It is characterized as a dimensionless value obtained by integrating the spectral UV irradiance within the wavelength range of 280–400 nm, adjusted using the erythemal action spectrum, and then multiplied by a scaling factor.

$$UVI = k_{er} \int_{280nm}^{400nm} E(\lambda) S_{er}(\lambda) d\lambda$$

where $E(\lambda)$ is the spectral irradiance, $S_{er}(\lambda)$ is the erythemal weighting function, and k_{er} is a constant. The UVI is a helpful tool for determining the health hazards associated with UV exposure and for issuing public health alerts [13].

2.2.2 Ultraviolet Protection Factor (UVPF)

A material's capacity to block ultraviolet (UV) radiation and shield human skin from damaging exposure is measured numerically by the ultraviolet protection factor (UVPF). When taking into account the biological efficacy of UV wavelengths, it is defined as the ratio of incident UV radiation to the amount transmitted through a substance. Greater protection is indicated by higher UVPF values; for instance, a UVPF of 50 indicates that just 2% of UV light enters the material [14].

UVPF evaluation is based on spectral transmittance measurements across the 280–400 nm ultraviolet region, encompassing both UVA and UVB wavelengths. These measurements are weighted using the erythemal action spectrum to account for the relative biological impact of different UV wavelengths. The protective efficiency of a material is influenced by multiple factors, including fiber composition, fabric construction, thickness, color, and applied chemical finishes. Owing to the health risks associated with excessive UV exposure, UVPF is widely adopted in international standards for the classification and labeling of UV-protective textiles [15]

3 Eco-Friendly Mechanisms for UV Protection

UVPF evaluation is based on spectral transmittance measurements across the 280–400 nm ultraviolet region, encompassing both UVA and UVB wavelengths. These measurements are weighted using the erythemal action spectrum to account for the relative biological impact of different UV wavelengths. The protective efficiency of a material is influenced by multiple factors, including fiber composition, fabric construction, thickness, color, and applied chemical finishes. Owing to the health risks associated with excessive UV exposure, UVPF is widely adopted in international standards for the classification and labeling of UV-protective textiles [15].

While helping produce vitamin D, ultraviolet (UV) light also induces oxidative stress, damages DNA and causes premature skin aging [16]. Sustainable UV-protective textiles decrease these risks by utilising eco-friendly fibers and functional finishes which absorb, reflect or scatter damaging UV photons without the use of environmentally dangerous chemicals providing lasting performance while ensuring material security and ecological integrity.

1. **Physical blocking**
2. **Chemical absorption**

3.1 Physical Blocking

Inherent structural features of textile materials such as fiber composition, use of filament vs staple yarns and fabric density govern physical blocking with respect to model optionally here. UV attenuation is determined by such factors and so not 100% transmittance can be achieved within these parameters. Without extensive chemical finishing, textiles can still obtain a high Ultraviolet Protection Factor UPF number by selecting eco-friendly fibers and optimizing the weave or knit structures [23]. That reconciles protective performance with environmental sustainability.

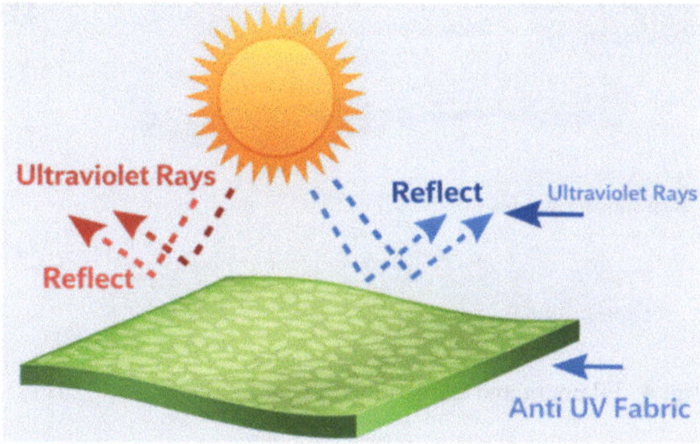

Fig. 3 UV radiation and fabric protection (Testex, 2024)

3.1.1 Fabric Structure on Functional Performance

Higher levels of UV protection in a textile are mainly attributed to the composition of its fabric. The weave or knit structure, yarn density, thickness and porosity of a fabric all directly influence its ability to block, absorb or deflect damaging UV rays. Fabrics with higher thread densities and tighter structures, with little or no space between yarns significantly decrease the transmission of UV radiation by reducing channels through which it can pass. Additionally, the type of yarn (filament vs. staple), yarn twist and fabric thickness all increase UV protection by complementing the physical barrier created between the skin and rays from the sun (Fig. 3).

3.1.2 Filament and Staple Fiber

The type of fiber (filament or staple) significantly determines the structure of yarn and fabric, specifically its porosity and cover factor. Although staple fibers yield bulkier, more porous fabrics with less innate protection, they produce denser and smoother fabrics that contain more UV protectant than do filament fibers [17]. Filaments Filaments are Long, continuous strands the continuity and smooth surface can make filaments pack closer together during yarn production. This tight packing creates a compact, homogenous yarn with very small fiber ends poking out. Hence, fabrics made up of filament yarns are low in bulk, not hairy much, and have flat surface. Because of the lesser irregularities on the surface, there is also reduced porosity which prevents more light and air from coming through. This results in denser fabrics compared with those made via filament, often improving UV protection quality. Because synthetic fibers, like polyester and nylon, are created with a smooth even diameter and shape. Filament fabrics are smooth and have a lustrous surface which tends to reflect UV radiation better than rougher staple fabrics (Fig. 4).

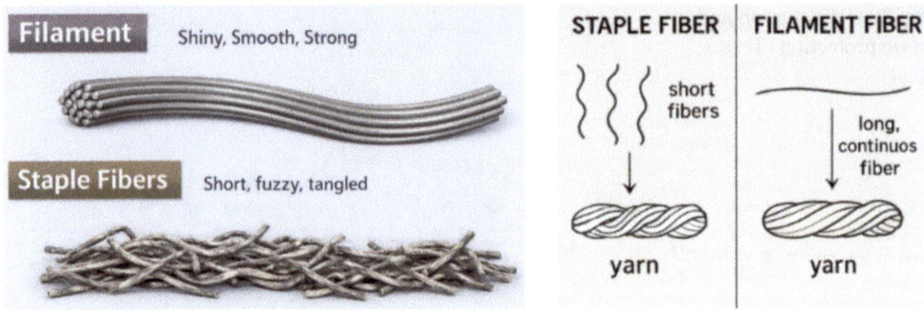

Fig. 4 Filament and staple Fiber Yarn (Textile School, 2011)

3.1.3 Yarn Twist

It is clear that yarn twist has significant effects on the characteristics of the fibers as well as their respective fabrics. The twist of yarn also affects the UV protection; as a rule, higher levels of twist increase UV protection due to increased fabric density, although in some cases low-twist yarns with high hairiness provide better protection, since more fibers scatter the UV light [18]. More twist = denser yarn and fabric A denser more structured fabric has a smaller porosity that reduces the amount of UV Light that passes through the fabric to skin. Compared to fabrics made with torque-free (ESTex- Elastic Stable Twist) yarns, ring-spun or twisted yarn garments typically have higher Ultraviolet Protection Factor UPF indicating that fabrics made up using a compacted form of spinning are more dense and therefore more screening following wet processing like laundry. Inversely, higher degrees of twist leads to less fiber protruding out from the yarn which can potentially reflect/absorb UV light, therefore less UV protection (Fig. 5).

Fig. 5 S twist and Z twist effect

3.1.4 Fabric Thickness

Thickness of fabric is a major factor affecting the UV protective ability of textiles. As a fabric thickens, more fibers work to absorb, scatter and reflect UV light, forming an increasingly stronger physical barrier. Generally, as fabric thickness increases, its mass per unit area and mass density increase while porosity decreases so that UV transmission is minimized and Ultraviolet Protection Factor (UPF) is maximized. This higher structural density extends the optical path length of UV light in the man-made piece, promoting more attenuation. Research shows that, in comparison to thinner competitors with otherwise similar construction parameters, thicker, higher-GSM fabrics consistently provide greater UV protection [19].

3.2 Chemical Absorption

Whereas, chemical UV absorption could take textile protection one step further, sustainable strategies demonstrate preference for natural and low processing compounds (e.g., chalcones, plant extracts, ZnO/ TiO_2 non-toxic nanoparticles), which achieve high UPF performance and enhances the safety in human health and environmental hazards. e.g. photostable and wash-durable formulations that not only minimize leaching of the chemical but also reduce their environmental effects, including organic UV absorbers e.g. benzophenones, triazine derivatives or small molecules—that should be used sparingly if at all.

1. Natural
2. Inorganic
3. Organic

3.2.1 Natural

Chalcones (1,3-diaryl-2-propen-1-ones) have been shown to absorb UV light, neutralize free radical, and possess anti-inflammatory activity which may work in concert as natural plant-based sunblock. Structures are conjugated for double-bonds, as seen amongst plant compounds like polyphenols, flavonoids, carotenoids and tannins which absorb UV photons and emit them in the form of harmless heat thus reducing the radiation that penetrates the skin [20]. These phytochemicals act as potent antioxidants that can neutralize reactive oxygen species created by UV exposure, averting oxidative damage of DNA, lipids, and collagen even in the face of some UV penetration. In addition, several plant extracts inhibit UV-induced pathways such as COX-2 and NF-κB that initiate redness, swelling, and cellular stress. Karanja and amaranth are two oils that strengthen the skin's lipid barrier and increase resistance to UV-induced dehydration.

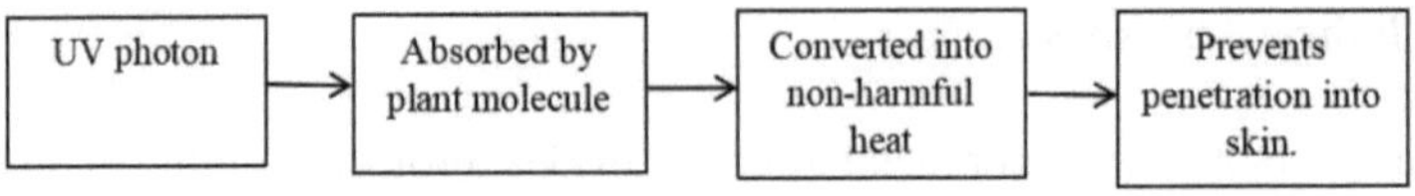

3.2.2 Inorganic

Titanium Dioxide

Titanium dioxide (TiO_2) creats a thin, inorganic coating on the fiber surface to improves the UV protection of textiles by absorbing, reflecting, and scattering ultraviolet light. TiO_2 particles, when applied as a nano or micro-sized finish, increase the surface area and optical density of the fabric, scattering incoming UV rays away from the textile and decreasing the amount that can pass through the pores in the yarn. At the same time, TiO_2 reduces UV transmission value through absorption of UV photons and convert them into innocuous energy with particular effect in regions of both UVA short wavelength and UVB. Photostable, non-yellowing and chemically inert titanium dioxide retains its UV-blocking ability even after repeated washing and photons exposure (of sunlight) [21]. The combination of physical shielding and photochemical absorption greatly increases the Ultraviolet Protection Factor (UPF) of TiO2finished fabrics making them suitable for technical textiles, outdoor wear, and protective apparel.

Zinc Oxide

Zinc oxide (ZnO) forms a long-lasting, mineral-based thin film on the surface of the fibers and has been shown to provide ultraviolet protection through absorption, scattering, and reflection which help improve UV protection in textiles.

This large band gap makes ZnO an effective UV absorber, especially in the UVA and UVB regions, which enables it to capture harmful photon energies in that range and convert them into heat [22]. When ZnO particles are applied in nanoscale or microscale, they stick to the surface of the textile and lengthen the UV light's optical path (Fig. 6). This extensive scattering lowers the amount of UV radiation that can pass through the yarn structure and fabric pores. ZnO also has antimicrobial and photocatalytic qualities that can enhance fabric hygiene and durability, though its photocatalytic activity is frequently regulated by surface modification to stop fiber deterioration. ZnO greatly raises the textile's Ultraviolet Protection Factor (UPF) because it is photostable, non-toxic, and resistant to discoloration. As a result, ZnO-finished fabrics are appropriate for outdoor clothing, protective gear, and technical textile applications.

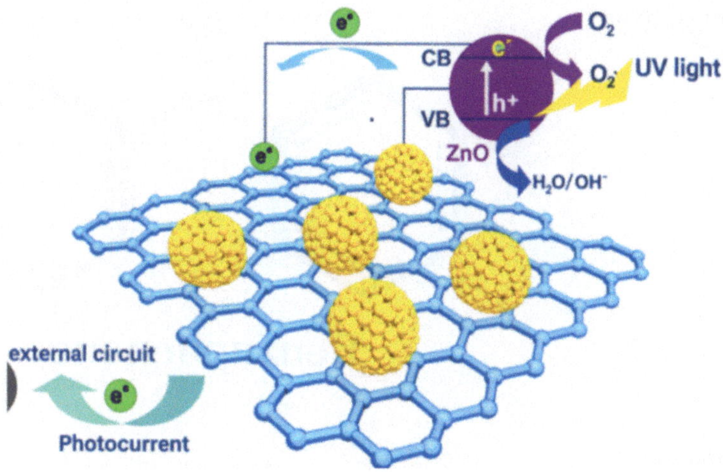

Fig. 6 ZnO-finished fabrics (**American Chemical Society** (ACS), 2012)

3.2.3 Organic Chemical

Benzophenones

By functioning as organic (chemical) UV absorbers that attach to or are embedded within fiber surfaces, benzophenones are Highly photostable, broad-spectrum UV absorbers improve the UV protection of textiles by selectively damaging UVA and UVB radiation. Conjugated aromatic rings and carbonyl groups in their molecular structure effectively absorb UV photons and transform the energy into innocuous vibrational heat, keeping UV light from penetrating the fabric. Benzophenones (Fig. 7) can be applied by padding, exhaustion, coating, or incorporation into polymer matrices because they are small, organic molecules that disperse evenly throughout the textile. Without appreciably changing the fabric handle or breathability, this deep fixation guarantees consistent UV-blocking performance. Furthermore, a lot of benzophenones are photostable, which means they don't break down in the presence of sunlight and continue to be protective even after repeated washings and wear. When used within a fabric, Benzophenones dramatically increase the Ultraviolet Protection Factor (UPF) of that material by absorbing UV radiation and dissipating it as thermal energy. They are commonly applied in apparel, outdoor textiles and technical fabric where you need something lightweight but still want it to be sun protective. It provides moderate environmental impact. They are synthetic, not biodegradable, and require careful handling to reduce environmental impact.

Fig. 7 Benzophenone (Shutterstock, 2026)

Benzotriazoles

Benzotrizoles are used as UV absorbers, primarily protecting textiles from harm in the UVA and UVB wavelengths. Composed of repeating triazole ring structures with powerful chromophores, the fabric absorbs energetic UV photons and dissipates them as heat via non-radiative decay, blocking pass-through of harmful UV rays to skin. As a textile finish, used on the basis of padding, coating or directly introducing it into polymer matrices (to form fibers), benzotriazoles get embedded in fiber or bonded on fiber surface to offer uniform protection without affecting fabric appearance or breathability. They also have a high degree of photostability, which means they do not degrade when exposed to sunlight over an extended period of time, and result in long-lasting UV protection relative to many other organic UV absorbers. Furthermore, the wash durability of benzotriazoles is often good if properly fixed / encapsulated, with the UPF of fabrics being retained after repeat use. The strong UV-absorption and energy-dissipation mechanism of benzotriazoles can greatly promote the UV-blocking efficiency of textiles applied in outdoor garments, sportswear and technical fabrics.

Hydroxyphenyltriazines

Hydroxyphenyltriazines(HPTs) possess excellent photostability and act as highly efficient, broad spectrum organic UV absorbers capable of protecting textiles from UV radiation. Their molecular structure—composed of a triazine core with hydroxyphenyl groups on its arms—forms robust chromophores capable of selectively capturing UVA and UVB photons and converting the absorbed energy into innocuous heat via ultrafast non-radiative relaxation. When directly applied to textiles via padding, coating, or incorporated into synthetic fiber dopes HPTs bind tightly onto the fiber surface or migrate into the polymer matrix and form a thin yet durable, transparent protective layer that does not change handle, color or breathability of fabric.

The hydroxyphenyltriazines offer better photostability, higher molar extinction coefficients and water resistance over older UV absorbers, such as benzophenones and ensure greater long-lasting protection in harsh sunlight. HPTs help increase the Ultraviolet Protection Factor (UPF) of fabrics in an efficient way; therefore are applied widely for high-performance outdoor textiles, sportwear and any technical fabric where durable, lightweight UV protection is needed.

Triazine-Based Absorbers

The strong chromophoric groups introduced in the triazine ring structure, triazine-based absorbers widen their protective capabilities against Ultraviolet radiation, acting as effective broad-spectrum Organic UV filters able of absorbing UVA and UVB wavelengths that protect the textiles from harmful effects. These molecules have high molar extinction coefficients, capable of capturing large amounts (99% out of UV) energy to itself and dissipating most into innocuous heat through fast non-radiative decay. Triazine-based absorbers, when applied to fabrics as a pad, coating or incorporation into spinning solutions of synthetic fibers for example, chemically bind around fiber surfaces or get trapped between polymer chains during preparation to create an invisible and permanent protective top layer that does not impact color. Handle or breathability. Supplemented with counterpart coatings, high-performance UV absorbers promise long-term photostability under direct sunlight with significantly lower degradation, delivering effective UV protection that remains intact for a longer lifespan than most of the first generation UV absorbers. They also demonstrate wash durability according to fixing/encapsulation, allowing them to be used repeatedly. This unique combination of efficient UV absorption, heat dissipation, and long-term stability allows triazine-based absorbers to make a significant contribution to increasing the Ultraviolet Protection Factor (UPF) of textiles used in outdoor clothing, sportswear and technical protective fabrics.

3.2.4 UV Absorbers Dyes (Both Inorganic and Organic)

UV protective dye and dye systems enhance the efficiency of ultraviolet protection in textile materials mainly via molecular absorption of UV radiation and energy dissipation mechanisms that minimize transmission of harmful rays through woven. Chromophores are structural components of conjugated polymers and aromatic molecules, which absorb light in ultraviolet (UV) range 280–400 nm causing absorption of the energy followed by conversion to heat radiation instead of transmittance through the textile so that they behave like UV absorbers embedded in the fibre matrix82225. In terms of UV protection, darker and more concentrated dye shades are expected to result in high extinction coefficients, leading to greater absorption of light from dyed textiles which increases the Ultraviolet Protection Factor (UPF) [24]. Studies show that both synthetic and natural colorants, when properly fixed on fibers, can greatly reduce UV transmittance. Related chemical groups and mordants further enhance the colorants' protective effect by improving fixation with spectral absorption. The additional combination of dyes with UV absorbers (derivatives of benzotriazole or benzophenone) when intrinsic UV-absorbing groups are incorporated also acts to further improve protection against UVA and UVB wavelengths. It demonstrates that the influence of dye chemistry on textile UV resistance is based on absorption, energy conversion and fixation mechanisms rather than purely physical as explained below in the flow chart.

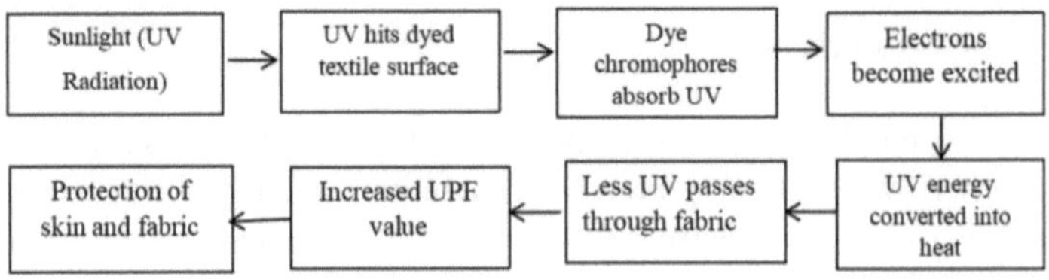

Table 2 enumerates a comparison study of different UV protection Compounds.

4 Impact of Different Textile Factors on UV Protection

The UV (Ultraviolet) protection offered by a fabric is affected by various factors such as fiber chemistry depending on composition, yarn structure such as cross- sectional shape of fibre, yarn twist, hairiness, types such as filament or spun yarn, fineness of yarn, and fabric geometrical parameters such as fabric construction, density, cover factor, porosity, thickness, and weight, etc. These factors influence how UV photons are absorbed, reflected, or scattered. In addition, environmental and usage conditions—such as moisture, stretching, aging, and garment design—can significantly alter the protection level. UV protection is determined by UPF (Ultraviolet Protection Factor). The higher the transmittance, the lower is the UPF and the lower is the UV protection factor and sustainability.

4.1 Effect of Fibre Chemistry

UV protection property of textile fabric is dependent on the types of polymer from which the fibre is made of. Natural and synthetic fibres show different UPF. Synthetic fibres offer more UPF than natural fibres [25] but they are not eco-friendly. Natural fibres are sustainable and among natural fibres, linen, viscose rayon, and white cotton have high transmission against ultraviolet ray and less protection abilities. Wool and cotton have a better UV protection effect than other natural fibres as their transmission is less. Existence of lignin makes the fibre more UV resistant. For example, bamboo and jute fibre offer strong UV protection due to the presence of lignin [26]. Moreover untreated cotton fabric has less transmittance than treated cotton fabric [27] and fabric dyed with dye or pigment boosts UV protection. Protein fibres (wool, silk) contain

Table 2 Comparison study of different UV protection Compounds [29]

Compound type	Description	Mechanism	Examples	Sustainability/ eco-friendliness
Natural	Plant-based UV absorbers	Absorb UV radiation naturally	Aloe vera, Cinnamomum camphora leaves, marigold flowers	High biodegradable, renewable, low toxicity; fully eco-friendly
Inorganic filters (Nanoparticles)	Stable metal oxides providing UV protection	Scatter and absorb UV radiation via electronic transitions	Titanium dioxide (TiO_2), Zinc oxide (ZnO)	Moderate durable, photostable, low toxicity; energy-intensive production; long-lasting but persistent in environment
Organic Filters	Organic molecules	Absorb UV radiation (UVA and UVB) and dissipate energy as heat or through isomerization	Benzophenone, Salicylic acid, Cinnamic acid, Triazine-based absorbers	Moderate synthetic, photostable, not biodegradable; environmental persistence possible; eco-conscious use when combined with physical barriers
UV-absorbing dyes (Inorganic & Organic)	Dye systems that absorb UV energy	Molecular absorption of UV radiation and energy dissipation	TiO_2/ ZnO-based pigments (inorganic), Synthetic organic dyes	Inorganic: Moderate–low toxicity, photostable; Organic: Low to Moderate–synthetic, may persist in environment; eco-friendlier if combined with natural absorbers

aromatic amino acids (e.g., tryptophan, tyrosine) which absorb UV radiation to some extent, giving moderate UV protection [28]. Among the synthetic fabrics the presence of benzene ring in the polymer structure makes the fibre inherently resistant to ultraviolet radiation. For example, polyester (PET), nylon, acrylic often have higher inherent UV resistance because of benzene rings or other UV-absorbing groups in their chemical structure. The UV absorption of polyester is below 310 nm which is exerted by phenyl ester group in the polymer chain [29]. Furthermore finishing the fabric with UV-absorbing nanoparticles such as TiO_2 or ZnO significantly enhance UV protection. For instance, coating cotton/PET twill fabric with TiO_2 nanoparticles raised UPF from ~ 3.7 to ~ 55.8 [30]. For producing sustainable untraviolate protective textiles, synthetic fibre should be avoided and they should be replaced by natural fibres.

4.2 *Effect of Yarn Structure*

UV protection property is also influenced by the yarn structure i.e., yarn twist, hairiness, fineness or coarseness, types i.e. filament or spun yarn and cross-sectional shape of fibre, etc.

4.2.1 Cross- Sectional Shape of Fibre

Non-circular fiber cross-sections (e.g., trilobal, flat, or hollow) can modify light reflection and absorption behavior. Such geometries increase internal scattering and reduce direct UV transmission through the yarn, thereby enhancing UPF (Wang et al. 2015). A yarn having triangular cross-sectional shape provides a higher UV protection effect whereas the lowest UV protection is provided from yarn with round shape [31].

4.2.2 Yarn Twist

Compactness and surface density of yarn is affected by yarn twist. Twist level is also affects yarn spacing. The higher the twist level, the closer the packing of fibres, the lower the inter-fibre spaces, the lower the UV transmission through the yarn and the higher the UPF factor [29].

4.2.3 Yarn Hairiness

UV protection of fabric is also affected by yarn hairiness. Hairy yarns, with protruding fibers on their surfaces, can increase UV scattering and absorption. These protruding fibers act as additional barriers and diffuse UV radiation, effectively extending the optical path within the textile (Sarkar, 2004). Thus, increased yarn hairiness enhances UPF.

4.2.4 Yarn Type

Type of yarn such as spun and filament yarn affect UV transmission. Spun yarns (from staple fibers) have a rougher, more irregular surface, creating multiple scattering sites for UV light, improving UV absorption whereas filament yarns, being smoother and more continuous, provide higher transmission and less scattering and lower UV protection. If they can form tightly contracted compact, low-porosity fabrics, they can exhibit good UV protection [32].

4.2.5 Yarn Fineness (Linear Density)

Yarn linear density determines the amount of material per unit length. Coarser yarns (higher tex or denier) result in denser fabrics with lower UV transmission because they contain more fiber mass to absorb, scatter radiation and more UV protection. Conversely, finer yarns produce more open structures that allow more UV rays to penetrate [33], less scatter radiation and less UV protection.

4.3 Fabric Geometrical Parameters

While chemical finishes and fiber type can modify UV blocking ability, **fabric constructional and geometrical parameters** significantly affect the degree of UV transmission and reflection [34].

4.3.1 Effect of Fabric Construction

UPF is affected by the fabric structure. The knit fabric has more open space than the woven fabric and for this reason UPF of knit fabric is less than the woven fabric which means lower ultraviolet protection ability [35]. The type of weave also affects the UV protection of fabric. The closer is the fabric's weave, the less free space in the fabric for UVR (Ultraviolet Radiation) to pass through to the skin because tightly woven fabrics takes the fibres more closer and the higher the protection against UVR and the higher the UPF value. In case of woven fabrics, higher UPF value is found in sateen fabric than twill and plain fabrics as sateen fabric provides higher cover factor and better protection against UVR [36]. Again, the satin fabric provides higher warp and weft density than plain and twill fabrics. The UPF value of plain woven fabrics (dyed) is at under 45 even with the tightest construction, while that of satin and twill weaves may be as high as 400.

4.3.2 Fabric Density, Cover Factor and Porosity

Density of fabric (ends and picks per inch in woven fabrics, or courses and wales per inch in knits) decides the tightness or looseness of fabric which determines how tightly the yarns are packed within a given area. Higher density corresponds to a higher cover factor, reducing inter-yarn gaps and, consequently, the UV transmittance. Densely woven fabrics with compact structures like twill and satin displayed higher UPF values than loosely woven plain fabrics [36]. Similarly, knitted fabrics with tighter stitch density show improved protection compared to open-knit designs. Tighter fabrics with higher cover and lower porosity offer lower UVR transmission capability and higher protection capability against UVR. Thus, with a decrease in porosity and an increase in cover factor, the UPF increases. However, the correlation between the cover factor and UPF was not linear and is depended on the colour of the fabrics also [29].

4.3.3 Weight and Thickness

Thickness is directly proportional to UV protection. Thicker fabrics absorb and scatter more UV radiation, leading to higher UPF values [37]. Several research showed that, a fabric's weight and thickness show a positive correlation with UPF values, which can be explained by the porosity. When weight per unit area is increases, the fabric porosity is decreased and cover factor is increased. In general, with an increase in weight and thickness of the fabric, the UPF increases, but this correlation is not linear [29, 37]. High cover factor is found in heavy, denser, and thicker fabrics showing blocking more ultraviolet radiation and transmitting less [29]. However, excessive thickness may hamper comfort (as greater thickness slows the movement of moisture vapor and liquid sweat through the material and moisture becomes trapped, increasing dampness and discomfort), breathability (because of restriction of airflow caused by more fibre), and thermal regulation (as thickness increases the amount of air held within the material, which raises thermal resistance and this reduces heat dissipation and can lead to overheating during activity) [38]. For clothing materials which are designed for hot climates, optimization between thickness and protection is required.

4.3.4 Effect of Dyeing on Ultraviolet (UV) Protection of Textiles

Dyeing has a considerable impact on the ultraviolet (UV) protective properties of textile materials because it changes the optical absorption properties and surface properties of textile materials. It is important to note that interaction between dye molecules and textile fibers cannot only give a color to the fabric but also influences the Ultraviolet Protection Factor (UPF) because it changes the capacity of the fabric to absorb, reflect, or transmit the UV radiation. A lot of dyes have chromophoric side chains that extend the absorption into the UV (290–400 nm), thus lowering the UV transmittance through the textile. Consequently, dyed fabrics tend to have higher values of UPF as compared to undyed or bleached ones, particularly those with darker dyes or high concentrations of dye. It is always noted that darker colors have better UV protection than light colors, as darker colors are able to block a wider spectrum of UV and visible light [37, 39]. As an example, Shiobara and co-workers [39] showed that cotton and polyester fabrics stained to darker colors had a significantly increased UPF value, but light pastel colors or light colors only provided a modest level of protection. Likewise, Sarkar [37] discovered that cotton fabrics dyed with natural colorants in deep colors had a higher improvement in UPF than light shades of the same dye, and it was established that UV protection is highly contingent on the depth of shade and level of dye concentration. Nevertheless, the enhancement of UV protection is not only a matter of the depth of color. Chemical structure and UV absorption spectrum of the dye are also important. Dyes with a large conjugation number, aromatic rings, or other auxochromic groups, e.g., reactive, vat, and natural dyes, can absorb UV radiation effectively, but some other dyes, such as sulfur dyes, can have weak UV absorption even though they look dark in color [39]. Moreover, dyeing may cause a chemical modification to the fabric, including decreasing porosity or fiber swelling or roughening the surface, all of which also lead to lower UV transmission. UV blocking can also be enhanced by the kind of mordants or auxiliaries used in dyeing, particularly when natural dyes are used, either via the generation of metal-dye complexes or by an increase in the scattering of light on the surface of the fiber. As an example, natural dyes such as madder, weld, or pomegranate peel with tannin or alum mordants have been used to reach UPF values of over 100 with the addition of durable UV protection [40]. However, UV-protective performance of dyeing may decrease with time under conditions of fading, washing, and photodegradation that decrease the concentration of dye and change the molecular structure. Thus, the longevity of colored fabrics in the sun and in the washing process is an important factor in the production of long-term UV-fighting clothing. Altogether, shading is an effective and sustainable method of augmenting UV protection to fabrics, which combines both aesthetic and functional functionality. Through careful choice of dye type, maximum shade depth, and fiber morphology, fabrics can be developed that have aesthetic color properties and effective and environmentally friendly UV protection.

4.4 Effect of Bleaching on Ultraviolet (UV) Protection of Textiles

Bleaching is one of the most widespread types of preparatory work in textile finishing, which is mainly aimed at removing natural pigments, waxes, and impurities of fibers to a uniform white outlook and preparing them to be dyeable. Bleaching, though, has a very negative occurrence of ultraviolet (UV) protection of textile materials. The process removes or degrades naturally occurring pigments and lignin-based materials that inherently absorb or scatter UV radiation, including the flavonoids, tannins, and carotenoids that are found in cotton, linen, silk, or other natural fibers. Accordingly, bleached fabrics have much lower Ultraviolet Protection Factor (UPF) values than unbleached fabrics or naturally pigmented fabrics [37, 41]. When these chromophoric substances are removed, the transmittance of the UV rays through the fabric also rises, making the fabric less able to block harmful UV-A (315–400 nm) and UV-B (280–315 nm) radiation. Bleaching also usually changes the physical structure of fibers, causing partial cellulose or protein chain degradation, porosity, and a smoother surface morphology, in addition to loss of pigments. These fabric alterations make the penetration of UV radiation easier, and the protective capability of the fabric is further reduced [42]. Indicatively, when cotton fabrics are bleached with hydrogen peroxide, the values of UPF have been reported to significantly decrease, and when such fabrics are bleached but not otherwise, the fabrics retain moderate

UV-blocking properties because such fabrics contain natural yellowish pigments [37]. On the same note, research on wool and silk has shown that their inherent UV absorption capability can be compromised due to large-scale bleaching, resulting in the destabilization of aromatic amino acid residues that serve to provide UV absorption. The degree of UV-protection loss varies according to a number of factors such as the nature and concentration of bleaching agents, the temperature of the process, the duration of exposure, and the composition of the fiber. Although an oxygen-based (e.g., H_2O_2) system is less damaging than the chlorine-based system, both can damage chromophoric structures that form the natural UV protection. Moreover, bleaching may be repeated or overbleached, causing oxidative damage, which causes microfibril disruption and subsequent reduction in UV resistance. Bleached white textiles, therefore being aesthetically valid, are more vulnerable to UV transmission and possible photodegradation when subjected to a long period under the sun. The bleaching of UV protection can be overcome, in practice, by a later step (e.g., dyeing with UV-absorbing colorants), a final finishing step (e.g., with organic UV absorbers), or a nanostructured coating (e.g., TiO_2 and ZnO). Such fabric modifications made post-bleaching are capable of restoring or even improving the UV-shielding capabilities in the fabric. However, in terms of sustainability and health-saving, it is important to have a perfect balance between whiteness and UV coverage. Consequently, factors that affect bleaching of fiber chemistry and optical behavior are essential in designing clothes with the ability to offer both the preferred aesthetics and a functional resistance to UV.

4.5 Effect of Stretching on Ultraviolet (UV) Protection of Textiles

The important mechanical factor that affects the ultraviolet (UV) protection properties of textile materials is stretching, particularly in knitted and elastic fabrics in sportswear, swimwear, and active outdoor clothing. The amount of stretch on a piece of fabric can significantly change the Ultraviolet Protection Factor (UPF) of the piece of fabric by changing the geometry, thickness, and porosity of the material on top of the fabric. Once a fabric is stretched, the yarns shift their positions, making the distance between fibers and yarn loops bigger. This architecture disfigurement results in an increase in porosity and a decrease in the fabric coverage that permits more UV radiation to travel through the open interstices [41, 43]. The UPF of a stretched piece of cloth, therefore, is often significantly smaller than the UPF of the same piece of cloth when relaxed. A number of studies have measured the adverse impact of stretching in terms of UV protection. Gambichler et al. [43] noted that for knitted polyester and nylon-elastane fabrics upon being stretched to 1030 percent of their original size, UPF reduced by above 50 percent, according to the extent of stretch. In the same way, [41] indicated that when the UPF of some stretchable sports fabrics showed 20% extension, the value dropped to less than 15 (moderate protection) as compared to 40 (excellent protection). The loss of the protection is attributed to the fact that, as the fabric is stretched, the optical density is reduced and the transmittance of the UV wave to the fabric is increased. In extreme cases, especially in lightweight or loosely knitted structures, the UV transmission can go as high as three times when the fabric is stretched past 30 percent [44]. Besides augmented porosity, stretching may also change the orientation of fibers and their refractive qualities, affecting the perception of light on the textile surface. The more the fibers are aligned under tension, the less surface scattering occurs with the UV rays, and more radiation makes it through. Mechanical deformation can also impact the microstructure of the synthetic polymers and can lead to a decrease in the inherent UV absorbance of the polymer chains in case the polymer chains experience orientation-induced crystallinity variations. Furthermore, stretching can also affect the disposition of dyes and UV-absorbing finishes on the fabric surface because when the coat layer is under tension, it will become thinner and less uniform, thus reducing the total UV-blocking performance [41, 42]. The degree of UPF degradation during stretch is dependent on a number of parameters, such as the type of fabric, weave/knit structure, composition of the fibers, linear density of the yarns, weight of fabric, and level of extension. TV knitted fabrics, because of their looped arrangement, are normally more vulnerable to stretch-induced UV transmission relative to woven fabrics. With lower to moderate stretch, fabrics with more elastane will be able to hold cover; however, repeated or prolonged stretching may still compromise UV performance. Thus, in applications where the fabric is subject to dynamic stretching, like in the case of swimwear or sportswear, fabric design needs to consider both sufficient baseline UPF values and high recovery to maintain UV protection over time. To reduce the negative effect of stretching, textile engineers commonly use the fabric constructions that are tighter, cover factors increased, use double knits, or have UV-absorbing finishes that can be used despite the deformation. However, a functional coating that incorporates inorganic nanoparticles (e.g., TiO_2 and ZnO) or organic UV stabilizers can be used to retain UV protection even when the coating is subjected to mechanical elongation. In a nutshell, stretching considerably undermines the UV protective properties of textiles by elevating porosity and density as well as lowering the optical density of the textile. Knowledge of mechanical strain and UV transmittance correlation is thus essential in designing and testing high-performance protective clothing, which can be used in the outdoor or sporting environment.

4.6 *Effect of Moisture and Wet Treatment on Ultraviolet (UV) Protection of Textile*

The effects of moisture and wet treatments on the ultraviolet (UV) protection effects of textile materials are far-reaching. The reaction of water and cloth changes various main physical and optical properties, including the light scattering, fiber refractive index, and porosity, which consequently changes the Ultraviolet Protection Factor

(UPF). Overall, as the textile gets wet as a result of perspiration, rain, or immersion in water, the UV protection levels go down sharply. This is of serious concern to fabrics finding application in outdoor, sports, or swimwear applications [41, 43]. When a fabric is wet, the inter-fiber and inter-yarn spaces are filled with water, and this decreases the air spaces, which would otherwise reflect incident UV rays. In fact, the refractive index of water and fibers are relatively similar, light scattering in the material becomes lower, and more UV is penetrated [44]. That is, water as an optical medium serves as the optical bridge enabling the UV rays to have an easier time going through the fabric. In addition, water tendencies cause the fibers to swell, creating a change in the geometry of the yarn and cloth, flattening the structure and decreasing the effective cover factor. On top of decreased scattering, structural distortion results in a devastating decrease in the value of UPF. The degree of UPF loss as a result of the presence of moisture is dependent on the type of fiber, fabric density, and the conditions of wetting. Hydrophilic fibers like cotton, viscose, and wool will absorb a lot of water, and thus the UV protection is lost more than synthetic hydrophobic fibers such as polyester or nylon [41]. As an illustration, Gambichler et al. [43] have illustrated that cotton fabrics had maximum reduction of UPF of up to 50 percent when they had been moistened, whereas polyester fabrics had only small percentage reductions. Also, lightweight or loosely made fabrics, which already possess reduced dry-state UPF, see more severe performance reductions in the wet state than do densely woven or knitted fabrics [44]. Humidness also affects the UV absorption of dyes and finishes that are added to the cloth. Water may change the chromophore environment, allowing some dye molecules or UV absorbers to be leached out, particularly in poorly fixed or low-substantivity systems. Likewise, the wet processing treatments (washing, scouring, or rinsing) may eliminate surface finishes that are superficial sources of UV blocking, resulting in long-term corrosion of protection [37]. The successive laundering cycles can also damage the fiber surface or weave, loosening it, and the UV transmittance also rises with time [41]. Fascinatingly, moisture can be a temporary protective factor against UV radiation in certain tightly woven or coated fabric since water can block the transmission in the short run [42]. Although, these are fabric-specific and not consistently safe in uniform photoprotection. As it applies to most real-world scenarios, especially with clothing that is subjected to sweat or wet natural conditions, moisture causes a significant reduction in UV-blocking functionality. In order to prevent such effects, the creation of hydrophobic surface finishes, moisture-wicking finishes, and long-lasting UV-absorbing finishes has become a significant field of textile engineering. NPs can be used in this case, like TiO_2 or ZnO, and the waterproof finishes can be water-repellent fluorocarbon finishes or silicone finishes to be used to ensure that UPF is maintained even in wet and humid environments. Thus, it is important to appreciate the influence of moisture and wet treatment on UV protection in the design of functional fabrics that maintain their protective properties in dynamic real-life settings.

4.7 Ecological and Human Health Impact of UV Protection Finishes

The use of ultraviolet (UV) protecting finishes on fabrics has become a growing trend to augment photoprotection, particularly when it comes to outdoor, medical, or sporting clothes. Nonetheless, though these functional finishes enhance the Ultraviolet Protection Factor (UPF) of textiles, there is the possibility of creating potential ecological and human health hazards owing to the chemicals and nanoparticles commonly used in their manufacturing. These finishes have an environmental impact based on their chemical makeup, method of application, the life cycle of the product, and subsequent release during the life cycle of the textile during their washing, wear, and disposal [45].

4.7.1 Ecological Implications

Many types of UV protection finishes are built on either inorganic nanoparticles (e.g., titanium dioxide (TiO_2) and zinc oxide (ZnO)) or organic UV absorbers (e.g., benzophenones, benzotriazoles, and hindered amine light stabilizers). The mechanisms by which these materials work are absorbing and scattering negative UV rays as well as reflecting the rays. Nevertheless, in the processing and laundering of textiles, nanoparticles or organic substances may be released into the wastewater systems, which may lead to ecotoxicity in water bodies [46]. It has been established that nano-TiO_2 and nano-ZnO released by treated textiles can induce oxidative stress in aquatic life and accumulate in the higher organism, interfering with cellular functioning and system homeostasis [47]. In addition, not all organic UV stabilizers are well biodegradable, and they might remain in the environment, hence causing environmental contamination in the long term. The treatment efficiency of these nanoparticles and chemical UV absorbers is not always complete in wastewater treatment plants. Consequently, effluents released due to textile industries or domestic washing may include a quantifiable amount of UV protection agents. These substances can react with natural organic matter or heavy metals, and this can increase their mobility and toxicity to the environment [49]. Also, nanomaterials used in UV protection have high energy usage and produce secondary pollutants, which adds to the environmental overall cost of the textile supply chain.

4.7.2 Human Health Concerns

Regarding the human health consideration, the toxicity, bioavailability, and route of exposure of active materials determine the safety of the UV protection finishes. Inorganic nanoparticles, especially TiO_2 and ZnO, can be deemed as photostable and seem to be less toxic in bulk. But their nanoscopic counterparts are raising concerns on skin penetration, inhalation, and systemic absorption [48]. Under specific conditions, e.g., long-term wear or contact with sweat, nanoparticles can be shed off fabric surfaces and contact the skin barrier. Though the majority of the research claims that penetration of the intact human skin is low, the damaged or sensitive skin may allow the uptake of nanoparticles in a limited form. Moreover, inhalation of the exposure during textile production, finishing, or wear by consumers (e.g., ironing or abrasion) is another possible risk, particularly for workers working with the nanosized powders or coating. The organic UV absorbers (e.g., benzophenone derivatives, e.g., oxybenzone) and benzotriazoles have been linked with endocrine disruption and skin sensitization in humans. Certain compounds may also imitate the presence of hormones, which results in reproduction and developmental toxicity with long-time exposure [49]. In addition, such materials can decay in sunlight to produce reactive intermediates that can produce free radicals or reactive oxygen species (ROS) that can either lead to skin irritation or fabric degradation. Thus, these finishes are better in terms of UV protection, but their safety is of critical importance over time.

4.7.3 To Sustainable UV Finishing Technologies

In order to eliminate ecological and health hazards, more attention is paid to the creation of environmentally friendly and bio-based UV protection finishes based on natural substances, including plant polyphenols, lignin, chitosan, or cellulose derivatives [50, 51]. Not only do these biopolymers have inherent UV absorption properties caused by their aromatic structures but also biodegradable properties, non-toxicity, and compatibility with sustainable textile processes. Moreover, surface modification methods that stabilize nanoparticles or entrap UV absorbers in the polymer mats can largely inhibit the leaching of nanoparticles during the washing process, which, in turn, will reduce the release of nanoparticles into the environment. The regulatory guidelines, including the European Union (REACH) (Registration, Evaluation, Authorisation and Restriction of Chemicals) and the U.S. EPA (Nanomaterial Stewardship Program), are taking a more active focus on risk assessment and disclosure of uses of nanomaterials in consumer products. The use of life cycle assessment (LCA) tools to assess UV protection finishes can allow improved assessment of its environmental and health effects, motivating the textile industry to use safer and more environmentally friendly ones. To conclude, although UV protection finishes are essential in enhancing the performance of fabrics and safety of consumers against solar radiation, it is necessary to design and use them with due care to their ecotoxicological and human health consequences. The move toward sustainable, bio-based, and non-toxic finishing systems is a vital aspect of finding a balance between functional performance and environmental responsibility in the area of textile innovation.

5 Conclusion

The World Health Organization (WHO) recommends photo protection by means of clothing for all categories of people. Although the textile fabrics provide excellent UV protection and low degradation occurs due to UV-radiation, UV protection properties have been influenced by composition of fiber, clothing design, fabric construction (open porosity, weight, and thickness), dyes, chemical agents or UV stabilizers applied to the fabric. However, UV protection is highly influenced by porosity or tightness of the fabric, and the absorption of UV radiation of fibers. Nowadays, there are some eco-friendly but expensive and complex methods for imparting UV protection capabilities to textile materials.

Functional textiles with UV-blocking effect have attracted numerous attentions for several decades. UV-blocking fabric could be fabricated via dry-pad-cure method, sol–gel method, hydrothermal method, flat screen technique and dip-coating method, self-assembly, etc. Generally, UV-blocking agents can be divided into two categories: organic and inorganic blockers. Organic blockers, also called UV absorbers, can absorb UV rays, such as benzophenone and natural dyes. Semiconductor oxides, such as titanium dioxide (TiO_2), zinc oxide (ZnO), silicon dioxide (SiO_2), etc., acted as inorganic UV blockers and exhibit remarkable UV-protection.

Green treatments in textile industry are getting more and more popular due to release of nanoparticles into the environment, human's body feels illnesses, and toxic effects. The competitive new technologies are striving to keep environment clean and are forcing industries to redesign their existing structures. Today, it is an obligation to be environmentally responsible in the textile and clothing industry. For this reason, green chemistry has attracted lots of attention recently.

Multifunctional textiles with different effects such as UV-blocking, antimicrobial, self-cleaning, and water-repellent can be obtained via green treatments of textiles. By using these treatments water and energy wastage can be reduced. Green treatments contain biodegradable, recyclable chemicals, which are not hazardous for both human and environment. The application of the principles of green chemistry and other aspects of clean technology will increasingly lead to more environmentally compatible manufacturing systems. Alike other industries, textile industry can continue to make its processes and chemistry green and provide a situation in which industrial growth and prosperity can be maximized without a negative impact on environment.

A variety of nanocomposites based on graphene and its derivatives have been scrutinized for UV protection applications due to constantly increasing global awareness on green/eco textile products. In this regard, sustainable UV-protective textile has been produced using natural polymers or biomolecules and plant extracts [52].

Beyond the synthetic UV-blocking agent materials' evident pollution of waters, they can also present environmental problems in the process of their production. According to the U.S. Environmental Protection Agency, this chemical accumulates in the gills of fish and can travel in high quantities up the food chain. In addition, synthetic materials are a significant challenge on land. Not only do these synthetic polymers slowly leach harmful chemicals in the soil, their longevity and non biodegradability are troublesome. Under these circumstances, we should spend a large amount of money for purifying domestic sewage for water reuse.

Material choice, nano-enabled finishes, and surface modification techniques have significantly enhanced the ultraviolet protection factor (UPF) of textile products as well as comfort, durability, and aesthetic appeal. Moreover, UV-rcsistant textiles are not limited to ordinary clothing, as their applications are increasingly found in sports, medical textiles, outdoor furniture, and work-related clothing. The importance of the field has also been added by the use of eco-friendly materials and sustainable production methods, which have introduced UV protection to the global sustainability goals. Nevertheless, the standardization of the testing procedures, ensuring the long-term viability of UV protection, and balancing the protective efficiency and wearability with affordability remain controversial, particularly as different industries may have varying standards and requirements for UV protection in their products. To sum up, UV-protective textiles are a fast-developing, diverse field of study and use. Textile scientists, material engineers, health experts, and industry stakeholders will have to remain collaborative to encourage innovation, enhance performance reliability, and increase customer awareness. The further development is expected to be focused on smart and multifunctional fabrics that will integrate UV protection with other characteristics and eventually enhance the well-being of human beings and make the interaction with the environment safer.

References

1. Ramirez R, Schneider J (2003) Practical guide to sun protection. Surg Clin 83(1):97–107
2. World health organization, & international commission on non-ionizing radiation protection (2002) Global solar UV index: a practical guide (No. WHO/SDE/OEH/02.2). World Health Organization
3. Tsuzuki T, Wang X (2010) Nanoparticle coatings for UV protective textiles. Res J Text Appar 14(2):9–20
4. Gasparro FP (2000) Sunscreens, skin photobiology, and skin cancer: the need for UVA protection and evaluation of efficacy. Environ Health Perspect 108(Suppl 1):71
5. Esen M, İlhan İ, Karaaslan M et al (2020) Investigation of electromagnetic and ultraviolet properties of nano-metal-coated textile surfaces. Appl Nanosci 10:551–561. https://doi.org/10.1007/s13204-019-01122-1
6. MacKie RM (2000) Effects of ultraviolet radiation on human health. Radiat Prot Dosimetry 91(1–3):15–18
7. Chinnasamy S, Ramachandran M, Sivaji C (2022) A study on ultraviolet radiation and its effects. REST J Adv Mech Eng 1(2):1–9
8. Hamouda SA, Alshawish NK, Abdalla YK, Ibrahim MK (2022) Ultraviolet radiation: Health risks and benefits. Saudi J Eng Technol 7(10):533–541
9. Ray A, Singha K, Pandit P, Maity S (2020) Advanced ultraviolet protective agents for textiles and clothing. In: Advances in functional and protective textiles, pp 243–260. Woodhead Publishing
10. Matsumura Y, Ananthaswamy HN (2004) Toxic effects of ultraviolet radiation on the skin. Toxicol Appl Pharmacol 195(3):298–308

11. Stankovic SB, Popovic D, Poparic GB, Bizjak M (2009) Ultraviolet protection factor of gray-state plain cotton knitted fabrics. Text Res J 79(11):1034–1042
12. Saravanan D (2007) UV protection textile materials. AUTEX Res J 7(1):53–62
13. Bashari A, Shakeri M, Shirvan AR (2019) UV-protective textiles. In: The impact and prospects of green chemistry for textile technology, pp 327–365. Woodhead Publishing
14. Gies HP, Roy CR, Elliott G, Zongli W (1994) Ultraviolet radiation protection factors for clothing. Health Phys 67(2):131–139
15. Gambichler T, Hatch KL, Avermaete A (2001) Influence of wetness on the ultraviolet protection factor of textiles. Br J Dermatol 144(3):495–499. https://doi.org/10.1046/j.1365-2133.2001.04090
16. Holick MF (2007) Vitamin D deficiency. N Engl J Med 357(3):266–281
17. Sarker P, Abedin MZ (2017) Influence of yarn and fabric structure on UV protection of cotton fabrics. J TextE Sci Eng 7(4)
18. Wong WY, Lam JKC, Kan CW, Postle R (2016) Impacts of yarn twist and staple length on UV protection of plain-knitted cotton fabrics. J Text Inst 107(12):1533–1542
19. Majumdar A, Das A, Hatua P (2015) Effects of fabric thickness and inter-yarn pore size on ultraviolet radiation protection by polyester woven fabrics. Fibers Polym 16:1163–1168
20. Milutinov J, Pavlović N, Ćirin D, Krstonošić MA, Krstonošić V (2024) The potential of natural compounds in UV protection products. Molecules 29(22):5409. https://doi.org/10.3390/molecules29225409
21. Trivedi M, Murase J (2017) Titanium dioxide in sunscreen', application of titanium dioxide. InTech. https://doi.org/10.5772/intechopen.68886
22. Zhou X-Q et al (2023) Zinc oxide nanoparticles: synthesis, characterization, modification, and applications in food and agriculture. Processes 11(4):1193. https://doi.org/10.3390/pr11041193
23. LeonService (2024) UV-blocking technology in clothing: how to shield your skin from UV damage? The importance of UV-Blocking clothing. Functional knit fabric wholesale manufacturer|SPORTINGTEX®. https://www.sportingtex.com/uv-blockingtechnology-inclothing#:~:text=UV%2Dblocking%20fabrics%20are%20designed,their%20integrity%20under%20various%20conditions
24. Saha B, Saha A, Das P, Kakati A, Banerjee A, Chattopadhyay P (2024) A comprehensive review of ultraviolet radiation and functionally modified textile fabric with special emphasis on UV protection. Heliyon 10(22):e40027. https://doi.org/10.1016/j.heliyon.2024.e40027
25. Buddhadeb et al (2024) A comprehensive review of ultraviolet radiation and functionally modified textile fabric with special emphasis on UV protection, Heliyon
26. Afrin T, Tsuzuki T, Wang X (2012) UV absorption property of bamboo. J Text Inst 103:394e399. https://doi.org/10.1080/00405000.2011.580543
27. Sarkar AK (2007) On the relationship between fabric processing and ultraviolet radiation transmission. Photodermatol Photoimmunol Photomed 23:191e196. https://doi.org/10.1111/j.1600-0781.2007.00306.x
28. Lee S et al (2021) Effects of ultraviolet light irradiation on silk fibroin films prepared under different conditions, Biomolecules
29. Saifur (2022) Protective textiles from natural resources, UV blocking protective textiles
30. Hadiseh et al (n.d.) UV protection properties of workwear fabrics coated with TiO_2 nano particle. Front Public Health, pp 1–10
31. Alicia et al (2021) Textile research journal, improving the ultraviolet protection factor of textiles through mechanical surface modification using calendering
32. Mukesh (2013) Journal of Textiles, Ultraviolet Protection by Fabric Engineering
33. Kan C-W (2015) PLOS-ONE. In: Vitro analysis of the effect of constructional parameters and dye class on the UV protection property of cotton knitted fabrics
34. Biswa (2010) The open textile journal, UV radiation protective clothing
35. Imran et al (2024) Journal of design and textiles, enhancing ultraviolet protection of bamboo fabrics using photochromic vat dyes

36. Polona et al (2023) Cotton woven fabrics as protective polymer materials against solar radiation in the range of 210–1200 nm
37. Sarkar AK (2004) An evaluation of UV protection imparted by cotton fabrics dyed with natural colorants. Biores Technol 92(3):301–307. https://doi.org/10.1016/j.biortech.2003.09.011
38. Slater K (1997) Comfort properties of textiles, juniper online journal material science, comfort properties of textiles: a review of some breakthroughs in recent resea
39. Shiobara T, Okabe Y, Shimizu Y (2009) The effects of dyeing on UV protection efficiency of apparel fabrics. Text Res J 79(2):123–130
40. Ultraviolet-protective textiles: exploring the potential of cotton knits dyed with natural dyes (2025) v 5(3):33. MDPI. https://doi.org/10.3390/materproc5030033
41. Gies P, Roy C, Toomey S (2009) The influence of fabric properties on the ultraviolet radiation protection provided by clothing. Photochem Photobiol 85(3):775–782
42. Algaba I, Riva A, Vialatte C (2013) Effect of bleaching processes on the UV protection of cellulosic fabrics. Cellul Chem Technol 47(3–4):227–234
43. Gambichler T, Hatch KL, Avermaete A, Altmeyer P, Hoffmann K (2002) Influence of wetness, stretching and periodic laundering on textile UV protection in vivo. Photodermatol Photoimmunol Photomed 18(3):135–140. https://doi.org/10.1034/j.1600-0781.2002.018003135.x
44. Lutkenhaus R, Sanz JR, Hatch KL (1999) The effect of stretch and moisture on the UV transmission of swimwear fabrics. Text Res J 69(2):98–103. https://doi.org/10.1177/004051759906900204
45. Štular D, Gorjanc M, Mozetič M (2020) Sustainable approaches to UV protective finishing of textiles. Materials 13(14):3170. https://doi.org/10.3390/ma13143170
46. Botelho MC, Costa C, Silva S, Costa S, Dhawan A, Fernandes A, Teixeira JP (2020) Nanomaterials in the textile industry: the need for sustainable innovations. Environ Res 182:109094. https://doi.org/10.1016/j.envres.2019.109094
47. Weir A, Westerhoff P, Fabricius L, Hristovski K, von Goetz N (2012) Titanium dioxide nanoparticles in food and personal care products. Environ Sci Technol 46(4):2242–2250. https://doi.org/10.1021/es204168d
48. Wang Y, Kalinina A, Menzies KL (2019) Human exposure to nanoparticles through textiles: a review. Front Bioeng Biotechnol 7:156. https://doi.org/10.3389/fbioe.2019.00156
49. Rimkus GG, Wolf M, Mindt-Prüfert S (2018) Occurrence and sources of benzotriazoles and benzophenone UV stabilizers in river waters in Germany. Environ Sci Pollut Res 25(4):3674–3683. https://doi.org/10.1007/s11356-017-0688-6
50. Grancarić AM, Tarbuk A, Botteri L (2017) Nanoparticles and their impact on textile finishing processes. Tekstil 66(7–8):225–238
51. Teli MD, Sheikh J (2013) Eco-friendly UV protective finishing of cotton using natural polyphenols from pomegranate peel extract. Fibers and Polymers 14(4):718–723. https://doi.org/10.1007/s12221-013-0718-9
52. Pandiyarasan V, Archana J, Pavithra A, Ashwin V, Navaneethan M, Hayakawa Y, Ikeda H (2017) Hydrothermal growth of reduced graphene oxide on cotton fabric for enhanced ultraviolet protection applications. Mater Lett 188:123–126

Eco-innovation in Textiles: Sustainable Technical Fibers and the Future of Fashion

Alba D'Aniello and Carmela Donato

Abstract The fashion industry is among the most resource-intensive and environmentally impactful sectors worldwide, due to high water consumption, chemical use, greenhouse gas emissions, and textile waste. These pressures have been further intensified by fast-fashion models characterized by accelerated production and consumption cycles. In response, sustainability and circular economy principles have become central to debates on the future of fashion, with fiber innovation emerging as a critical lever for reducing environmental impact. This chapter examines technical eco-innovation in textiles by focusing on sustainable and innovative fibers as a pathway toward a more circular and responsible fashion system. Drawing on the literature on sustainable fashion, circular economy, and green innovation, the chapter explores how fibers derived from waste and renewable biological sources can contribute to waste reduction, resource efficiency, and value creation. Particular attention is devoted to the luxury fashion sector, which faces both opportunities and tensions in integrating sustainability while preserving quality and exclusivity. Based on the results from a narrative review and multiple case-based analysis (Orange Fiber, Piñatex, and Mylo) the chapter illustrates different models of sustainable fiber innovation, including waste-to-resource processes, plant-based leather alternatives, and biotechnology-driven materials. The analysis highlight both the potential and the challenges of scaling sustainable fibers, emphasizing the importance of supply-chain collaboration, technological maturity, and consumer acceptance. Overall, the chapter positions sustainable technical fibers at the intersection of innovation, circularity, and consumer culture, contributing to the rethinking of the future of fashion.

Keywords Sustainable fashion · Circular economy · Material innovation · Sustainable technical fibers · Sustainable innovation · Luxury fashion sustainability · Case-based analysis

A. D'Aniello (✉)
Adjunct Faculty, Luiss Guido Carli, Rome, Italy
e-mail: adaniello@luiss.it

C. Donato
Department of Economics, University of Roma Tre, Rome, Italy
e-mail: carmela.donato@uniroma3.it

S. S. Muthu (ed.), *Technical Textiles for Sustainability*, Sustainable Textiles: Production, Processing, Manufacturing & Chemistry, https://doi.org/10.1007/978-3-032-29974-1_6

1 Introduction

The fashion industry is widely recognized as one of the most resource-intensive and environmentally impactful sectors worldwide, due to its extensive water consumption, intensive use of chemicals, and growing volumes of textile waste [28, 59]. These pressures are particularly pronounced in the context of fast fashion, where accelerated production and consumption cycles have intensified both environmental degradation and social concerns [33]. As demand for fashion products continues to rise, the cumulative effects of material extraction, manufacturing processes, and end-of-life disposal pose significant challenges to the long-term sustainability of the industry.

Within this context, innovation has emerged as a central mechanism for mitigating fashion's environmental footprint and enabling the transition toward more sustainable and circular systems. Among the various forms of innovation, textile fibers play a particularly critical role, as materials account for a substantial share of the industry's ecological impact. In fact, textiles represent the fifth-largest area in terms of primary raw material use within the European economy (European Commission, EU Strategy for Sustainable and Circular Textiles). Prior research highlights how the redesign of materials and production processes can support alternative business models capable of reducing resource use while preserving value creation [48, 76]. In this sense, sustainable fibers offer the opportunity to address environmental challenges at their source, by embedding sustainability directly into the material composition of fashion products rather than treating it as a downstream adjustment.

This chapter examines the transformative potential of technical sustainable fibers in reshaping the textile and fashion industry. Focusing on three emblematic case studies—namely, Orange Fiber, Piñatex, and Mylo—the chapter illustrates how fibers derived from citrus waste, pineapple leaves, and mycelium can generate both ecological and economic value. These innovations exemplify different approaches to material development that foster cross-industry collaboration and contribute to the valorization of agricultural and food waste streams, aligning environmental objectives with industrial feasibility and market needs [41].

Beyond technological and production-side considerations, the transition toward sustainable fibers is also shaped by consumer perceptions and acceptance, particularly among younger generations. Prior research suggests that younger consumers exhibit growing sensitivity to environmental and ethical issues and increasingly integrate sustainability considerations into their fashion preferences [36, 56]. The analysis highlights a generally favorable orientation toward innovative sustainable fibers, associated with reduced environmental impact and ethical benefits, while simultaneously revealing recurring concerns related to durability, performance, price, and limited familiarity. Such perceptions underscore the dual nature of material innovation: novelty can enhance interest and perceived innovativeness but may also generate uncertainty when materials are insufficiently understood or communicated.

By integrating case-based evidence with illustrative consumer insights, the chapter adopts a holistic perspective on sustainable fiber innovation, capturing both supply-side dynamics and demand-side perceptions. The analysis ultimately positions sustainable fibers at the intersection of innovation, circular economy principles, and consumer culture, offering a structured discussion of the opportunities and tensions that characterize their adoption, particularly within the luxury fashion sector. In doing so, the chapter contributes to ongoing debates on how eco-innovation in textiles can move beyond experimentation and toward scalable, credible, and value-creating solutions for the future of fashion.

This chapter begins with a review of the literature on sustainable fashion, focusing on its opportunities and challenges. It first examines the environmental costs of the fashion industry, and then discusses green innovation, with particular attention to economic trade-offs and the role of the luxury sector.

The chapter then explores innovation for sustainable fashion, highlighting the role of technical innovative fibers in reducing environmental impact.

Next, a multiple case study analysis is presented, focusing on Orange Fiber, Piñatex, and Mylo, followed by a cross-case comparison to identify key similarities and differences.

Finally, the discussion section outlines the main implications, emphasizing both the potential and the challenges of scaling technical sustainable fiber innovations.

2 Sustainable Fashion: Opportunities and Challenges

2.1 The Environmental Costs of Fashion

The fashion industry exerts significant pressure on the environment across its entire value chain, from raw material extraction and fiber production to manufacturing, distribution, use, and disposal. At the production stage, fashion relies heavily on natural resources such as water and land (for example, cotton needs is extremely water-intensive) [25, 84]. Furthermore, the fashion industry is responsible for nearly 10% of greenhouse gas emissions and 20% of water waste, while consuming a significant amount of energy resources (i.e., more than aviation and shipping combined) [21]. The manufacturing processes (particularly dyeing and finishing) frequently involve hazardous chemicals that pose risks to ecosystems and human health [19, 32]. The widespread use of synthetic fibers derived from petrochemicals, such as polyester and nylon, further contributes to pollution through greenhouse gas emissions, toxic effluents, and microplastic release.

These impacts are especially pronounced in the context of fast fashion, where accelerated production and consumption cycles amplify environmental degradation and social externalities [33, 59]. Although estimates regarding fashion's contribution to global emissions and resource use vary across studies, there is broad consensus that the sector represents one of the most environmentally intensive consumer industries, particularly in terms of water use, chemical pollution, and textile waste [21, 28]. The growth trajectory of the industry further exacerbates these challenges: global textile production nearly doubled between 2000 and 2015, and clothing and footwear consumption is projected to increase by more than 60% by 2030 [30].

In response to these mounting pressures, policymakers and industry actors have increasingly questioned the sustainability of the traditional linear "take–make–waste" model. At the European level, the EU Strategy for Sustainable and Circular Textiles emphasize durability, recyclability, and waste reduction as key priorities for the sector's transformation [30]. Similarly, international policy forums and industry platforms have placed fashion sustainability at the center of climate and development debates [37]. However, despite growing awareness and reporting efforts, industry performance remains uneven, revealing a persistent gap between sustainability commitments and material outcomes. Indeed, recent industry commentary suggests a growing tension between sustainability narratives and public trust, particularly in luxury, where reputational crises and alleged misconduct have contributed to a more skeptical climate toward "sustainable fashion" claims. In this context, sustainability communication risks being perceived as performative unless supported by material and verifiable changes. This credibility challenge reinforces the relevance of focusing on fiber innovation as a concrete, upstream lever for environmental impact reduction [75].

Within this context, the circular economy has emerged as a dominant framework for rethinking fashion systems. By replacing end-of-life logic with restorative processes based on renewable inputs, waste reduction, and the elimination of toxic chemistry, circular models aim to decouple value creation from resource depletion [27]. Practices such as recycling, upcycling, repair, and remanufacturing are increasingly promoted as mechanisms to extend product lifecycles and reduce environmental burdens [11, 39]. Yet, their effectiveness depends on technological feasibility, supply-chain coordination, and consumer acceptance—issues that remain particularly complex in fragmented global fashion systems.

2.2 *Green Innovation: Economic Trade-Offs and the Role of Luxury*

Beyond regulatory and systemic shifts, sustainability in fashion is increasingly framed as an innovation challenge. Green innovation—defined as the development of products, processes, and business models that simultaneously improve environmental and economic performance—has been shown to generate competitive advantages through waste reduction, lower resource consumption, and reputational benefits [51, 57]. From a managerial perspective, sustainability can stimulate learning, foster product differentiation, and support long-term value creation [6, 79].

However, the relationship between sustainability initiatives and economic performance is not automatic. Empirical evidence suggests that certifications, environmental management systems (e.g., ISO 14001), and responsible sourcing practices may produce mixed results, particularly when firms struggle to coordinate complex supply chains or absorb higher costs [12]. These trade-offs are especially salient in the fashion industry, where cost pressures, speed-to-market, and quality expectations often conflict with sustainability goals.

The luxury sector occupies an ambivalent position within this debate. On the one hand, luxury goods are often characterized by superior quality, durability, and longer lifecycles, which may render them inherently more sustainable than mass-market alternatives [4, 60]. On the other hand, research indicates that sustainability cues can sometimes undermine perceived luxury quality and purchase intentions, creating a tension between traditional luxury codes and environmental signaling [1]. At the same time, younger consumers increasingly expect luxury brands to uphold higher ethical and environmental standards, and sustainability-oriented communication has been shown to reposition luxury as simultaneously exclusive and responsible—"gold and green" [5, 9].

These dynamics suggest that sustainability in fashion, and especially in luxury, cannot rely solely on symbolic communication or downstream practices. Instead, it requires interventions that address the material foundations of products, where a substantial share of environmental impact is concentrated.

3 Innovation for Sustainable Fashion

Innovation is widely recognized as a key enabler of sustainability transitions, yet in fashion its implementation is often constrained by structural and operational bottlenecks. In particular, the adoption of sustainable innovations remains limited by weaknesses in the collection, sorting, and preprocessing infrastructure and by the difficulty of ensuring consistent output quality at scale [7]. These constraints are especially critical for fashion value chains, where speed-to-market, cost discipline, and stringent quality standards coexist with increasing expectations for environmental and social responsibility.

Importantly, the early phases of sustainable innovation are characterized by uncertainty, which can affect both firm decisions and consumer responses. Companies therefore need to rely on partnerships to reduce technological and market risk, while simultaneously ensuring that product usability and performance remain credible to end users [23, 73]. In this regard, communication plays a functional—not merely symbolic—role: by highlighting concrete product attributes and consumer-relevant benefits, firms can strengthen perceived innovativeness and lower perceived risk, facilitating acceptance of novel solutions [40].

As markets face growing volatility and environmental pressures, fashion companies and their partners are also adjusting supply-chain structures and business models to accommodate emerging requirements, including resilience and adaptability [42].

Against this backdrop, the present chapter focuses on material innovation as a high-leverage pathway for sustainability in fashion: rather than treating sustainability as a downstream adjustment, it considers how innovation at the fiber level can shape product performance, scalability, and consumer acceptance—thereby creating the conditions for diffusion beyond niche experimentation.

3.1 Technical Innovative Fibers for Fashion

Technical innovation in textile fibers has increasingly emerged as a central pathway for advancing sustainability in fashion, as fibers determine a substantial share of environmental impacts related to resource use, emissions, durability, and end-of-life options [32, 34, 59]. From a material perspective, the fashion industry traditionally relies on a limited set of dominant fibers—cotton, polyester, viscose, and wool—which together account for most of the global textile production. While these fibers differ in origin and performance, all are associated with significant environmental externalities, including intensive water use, chemical inputs, greenhouse gas emissions, and microfiber pollution [19, 35].

Building on the distinction between natural fibers and processed fibers, the latter including synthetic and regenerated materials, prior research highlights how fiber type, processing intensity, and finishing treatments critically shape environmental outcomes [32, 69]. To clarify the main categories of textile fibers relevant to sustainability and innovation, Fig. 1 provides a simplified taxonomy of fibers used in the fashion industry. Synthetic fibers such as polyester and nylon—derived from fossil fuels—are particularly problematic due to their contribution to microplastic pollution, with textile fibers have been frequently linked to a substantial share of microplastics released into aquatic environments (e.g., [7, 35]). At the same time, natural fibers such as cotton and wool are not impact-free, especially when intensive agriculture, chemical dyes, and finishing processes are involved [59].

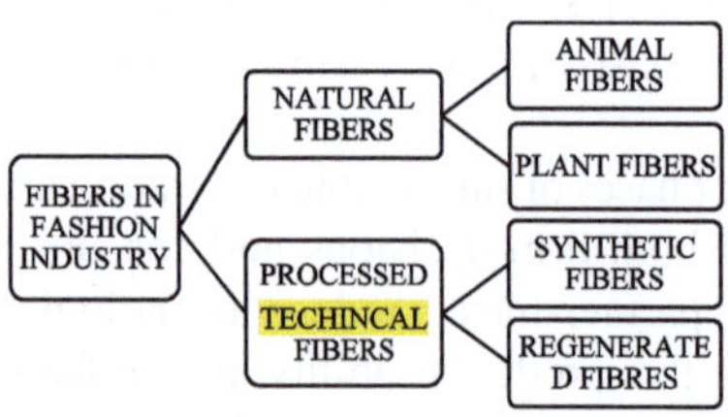

Fig. 1 Fibers for fashion. *Source* Authors' elaboration. The figure distinguishes between natural fibers (animal- and plant-based) and processed fibers, including synthetic and regenerated fibers. This classification provides a conceptual framework for understanding environmental impacts, material properties, and innovation pathways in textile production

In response to these challenges, technical innovation in fibers has followed multiple trajectories. One stream focuses on improving existing fibers through recycling and regeneration technologies. Advances in mechanical and chemical recycling have enabled the development of regenerated fibers such as Econyl®, which can be recycled repeatedly without loss of quality and are increasingly adopted by fashion brands [29, 81]. Another stream centers on bio-based and agro-waste-derived fibers, which transform agricultural and food waste into textile inputs, thereby reducing reliance on virgin resources and contributing to circular value creation [3, 65, 66, 70]. Recent studies demonstrate the technical feasibility of producing cellulose fibers from citrus waste, pineapple leaves, banana fibers, and other agro-industrial residues, opening new avenues for sustainable yarns and fabrics [44, 47, 68].

A third and more experimental trajectory involves biofabrication and biomaterials, where fibers or leather-like materials are grown or engineered through biological processes, such as mycelium-based materials [82]. These innovations challenge traditional boundaries between natural and synthetic fibers, raising new questions about scalability, cost, performance consistency, and consumer acceptance. While industry reports highlight growing investment and experimentation in this area, they also point to persistent barriers related to technological maturity and industrial integration [7, 55].

From a circular economy perspective, fiber innovation intersects with practices such as recycling and upcycling, which differ both technically and perceptually. Recycling involves transforming waste materials into new fibers of equivalent value, whereas upcycling repurposes discarded materials into products with higher perceived value [2, 45, 63]. Consumer research suggests that upcycled products may benefit from associations with creativity, uniqueness, and novelty, which can enhance product evaluations and purchase intentions [2, 18, 80]. However, sustainability cues can also generate trade-offs: recycled or eco-friendly materials may trigger concerns about quality or performance, particularly in luxury contexts [1, 53].

These tensions are especially salient in the luxury sector. On the one hand, luxury goods are often characterized by durability, craftsmanship, and longer life-cycles, which can render them inherently more sustainable than mass-market alternatives [4, 60]. On the other hand, sustainability signals may conflict with traditional luxury codes if they are perceived as compromising quality or exclusivity [1]. Research shows that consumers react negatively to unsustainable luxury when ethical expectations are violated, generating guilt and negative word-of-mouth [5], while sustainability-oriented communication can successfully reposition luxury as both premium and responsible when supported by credible material innovation ("gold and green"; [9]).

Importantly, consumer acceptance of innovative fibers depends not only on environmental benefits but also on novelty, familiarity, and perceived usability. Novel attributes can enhance product evaluations by signaling innovation and differentiation [20, 58], yet excessive unfamiliarity may increase perceived risk, particularly for materials worn close to the skin. Studies on sustainable luxury indicate that consumers' responses are shaped by a combination of material properties, aesthetic cues, and personal values, including environmental concern and decision-making styles [43, 71].

Overall, the literature suggests that innovative fibers represent a high-leverage intervention for sustainability transitions in fashion, as they operate upstream at the level of material choice while simultaneously influencing downstream perceptions and behaviors. However, their diffusion depends on the alignment of technological feasibility, supply-chain integration, and consumer meaning-making. This perspective provides the conceptual foundation for the following case studies, which examine how Orange Fiber, Piñatex, and Mylo operationalize different pathways of fiber innovation—respectively waste-to-cellulose, agro-residue valorization, and biofabrication—within real market and branding contexts.

4 Case Studies of Sustainable Fibers in the Fashion Industry

To examine how sustainable fiber innovation is operationalized in the fashion industry, this chapter adopts a qualitative multiple case study approach [26, 83]. Case studies are particularly suitable for exploring complex, emergent, and context-dependent phenomena, such as material innovation at the intersection of sustainability, technology, and market adoption, where clear boundaries between technological processes, organizational strategies, and consumer meanings are difficult to disentangle.

The analysis focuses on three information-rich cases—Orange Fiber, Piñatex, and Mylo—selected through theoretical and purposive sampling rather than representativeness.

The cases were chosen because:

1. they introduce innovative fibers explicitly framed as sustainable or circular,
2. they rely on distinct innovation logics (waste-to-cellulose, agro-residue valorization, biofabrication),
3. they have achieved visibility within the fashion industry, including collaborations with established brands, and
4. they offer extensive publicly available documentation, enabling systematic qualitative analysis.

Data collection is based exclusively on secondary documentary sources, consistent with prior research on sustainability-oriented innovation and luxury industries. For each case, we analyzed multiple types of materials, including: (a) official company websites and material descriptions; (b) sustainability, impact, or technical reports; (c) press releases, interviews, and media coverage; (d) certifications, partnerships, and industry affiliations; and (e) textual and visual materials describing production processes, material properties, and brand narratives. Sources were selected to capture both technological aspects (feedstock, production processes, material performance) and strategic framing (sustainability narratives, market positioning, partnerships).

The analysis followed an iterative thematic approach. First, each case was examined independently to identify key dimensions related to feedstock origin, production logic, material characteristics, industrial collaboration, and sustainability framing. Second, a cross-case comparison was conducted to highlight similarities and differences across innovation trajectories, degrees of technological maturity, and market integration. This process allowed us to connect empirical observations with the conceptual lenses introduced earlier—namely circular economy, green innovation, and sustainable luxury—while avoiding claims of generalizability beyond the illustrative scope of the cases.

4.1 Orange Fiber: From Citrus Waste to Technical Innovative Textiles

A growing body of research highlights the potential of agri-food waste—such as fruit peels, pulp, and seeds—as an underutilized resource for circular applications across multiple sectors, including fashion [3, 66, 68, 70, 82]. Orange Fiber represents one of the most emblematic Italian examples of this waste-to-resource trajectory applied to technical textile innovation. Founded in Catania in 2012 by Enrica Arena and Adriana Santonocito, Orange Fiber was conceived with the explicit goal of valorizing citrus-processing by-products and, at the same time, promoting a vision of Sicily as a territory capable of combining sustainability, innovation, and textile tradition [10, 62].

The company addresses a structurally relevant problem within the Italian agri-food system. Citrus juice production generates large volumes of residual biomass composed of peels, pulp, and seeds that are no longer usable for food purposes. In Italy, more than 700,000 tons of citrus by-products are produced annually, and approximately 60% of the weight of a squeezed orange becomes waste, traditionally associated with disposal costs and environmental burdens [82]. Orange Fiber's core innovation lies in reinterpreting this residual material as a valuable input for textile production rather than as an end-of-life problem.

The fiber obtained from citrus residues is cellulose-based and can be engineered to meet the aesthetic and functional requirements of premium apparel. Depending on processing choices, the resulting fabrics can exhibit a soft, silk-like hand feel, lightness, breathability, and a glossy or matte appearance, making them suitable for high-end fashion applications. Importantly, the material can be used either in its pure form or blended with other fibers, allowing designers and manufacturers to tailor performance and aesthetics to specific product categories [82].

A distinctive feature of Orange Fiber's development path is its reliance on industrial and research partnerships to move from laboratory experimentation to market applications. Early collaboration with the Politecnico di Milano played a key role in the development of the patented production process. Subsequently, Orange Fiber established partnerships with established fashion and textile players.

A notable example is the collaboration with Salvatore Ferragamo, which led to an exclusive agreement and the launch of a women's collection in 2017, significantly increasing the material's visibility within the luxury fashion system [62]. The fiber was later adopted by other brands, including H&M within its Conscious Collection, demonstrating its potential compatibility with both luxury and large-scale fashion contexts.

Another strategic collaboration involves the Lenzing Group, a global leader in specialty fibers. Together, the partners developed a lyocell fiber marketed under the TENCEL™ brand, combining wood pulp with citrus-derived cellulose. This partnership illustrates a scale-oriented pathway in which sustainable innovation is integrated into existing industrial infrastructures rather than remaining confined to niche experimentation [14, 62, 77].

From a sustainability perspective, Orange Fiber frames its business model within circular economy principles, emphasizing waste valorization, resource efficiency, and process optimization. The company has articulated medium- and long-term sustainability objectives, including the transition toward circular management of water, energy, and resources by 2030, as well as the reduction of deforestation impacts by avoiding the equivalent of approximately 5000 trees per year by 2026 [62]. These goals are supported by incremental process innovations aimed at reducing water and energy consumption, such as investments in new machinery and alternative drying techniques [77].

Despite its promise, the Orange Fiber case also illustrates typical challenges associated with bio-based and waste-derived fibers. Transforming organic residues into consistent, high-quality textile outputs remains technologically demanding and potentially costly. Scaling depends on stable access to citrus waste streams, effective logistics integration, and reliable industrial partners capable of ensuring quality consistency. Moreover, because the fiber is intended for garments worn close to the skin, product safety and certification play a strategic role. In this respect, the adoption of OEKO-TEX® Standard certification is positioned as a key credibility mechanism supporting consumer trust and regulatory compliance [62].

Overall, Orange Fiber exemplifies how agri-food waste can be reintegrated into high-value textile supply chains, aligning material innovation, industrial feasibility, and sustainability narratives within both luxury and mainstream fashion contexts.

Production Process and Circular Logic

Orange Fiber's production process is based on the extraction of cellulose from citrus pastazzo, the residual biomass generated during industrial juice production. A key design choice is the proximity between juice-processing facilities and fiber extraction sites, which reduces transportation needs and related environmental impacts [62, 72].

The technical process involves a sequence of mechanical and chemical treatments aimed at isolating and purifying cellulose suitable for textile use. While some stages may involve chemical agents—such as hydrogen peroxide—the overall system is designed to deliver net environmental benefits compared to conventional waste disposal and virgin fiber production, with reported reductions in pollution levels relative to traditional alternatives [72, 82].

Once extracted, the cellulose is converted into a spinnable material and transformed into yarn through processes compatible with existing textile infrastructures. Yarn spinning may involve specialized international partners, while final weaving and finishing typically take place in Italy, reinforcing the connection with domestic textile manufacturing and craftsmanship [77].

The entire process is protected by patents first registered in Italy in 2014 and subsequently extended to other major citrus-producing countries. From a circular economy perspective, the production model exemplifies upcycling: a low-value by-product is transformed into a high-quality textile input, reducing waste while generating economic and symbolic value. This integration of material innovation, industrial partnerships, and circular logic positions Orange Fiber as a reference case for waste-based textile innovation in fashion.

Brand Collaborations

Brand collaborations have played a crucial role in the legitimation and diffusion of Orange Fiber within the fashion system, particularly by translating material innovation into recognizable and credible market offerings. Rather than relying on direct-to-consumer strategies, Orange Fiber has pursued a partnership-based model in which established fashion brands act as intermediaries between technological novelty and consumer acceptance. This approach reflects a broader pattern observed in sustainable material innovation, where collaboration with reputable brands reduces perceived risk and enhances legitimacy, especially in premium and luxury contexts.

One of the most significant collaborations was established with Salvatore Ferragamo, which entered into an exclusive agreement with Orange Fiber in 2016. The partnership resulted in the launch of a women's collection in 2017 featuring garments made with citrus-based textiles. This collaboration was strategically relevant for two reasons. First, it positioned Orange Fiber within the luxury segment, signaling that waste-derived materials can meet high standards of quality, aesthetics, and performance. Second, Ferragamo's strong association with Italian craftsmanship and heritage helped frame material innovation as an extension of tradition rather than a rupture, thereby reducing potential resistance linked to novelty.

Beyond luxury, Orange Fiber has also collaborated with H&M for the brand's Conscious Collection. This partnership demonstrates the versatility of the material and its potential scalability across different market segments. While the sustainability positioning of fast fashion brands has been widely debated, the rapid sell-out of the Orange Fiber-based garments suggests that material innovation, when embedded within familiar brand architectures, can attract consumer interest even in high-volume contexts. Importantly, this collaboration illustrates a different diffusion logic compared to luxury partnerships: here, the emphasis shifts from exclusivity and craftsmanship to accessibility and visibility.

In addition to fashion brands, Orange Fiber's collaboration with Lenzing Group represents a critical step toward industrial scalability. The joint development of a citrus-infused lyocell fiber under the TENCEL™ brand integrates Orange Fiber's innovation into an established specialty-fiber ecosystem. This partnership highlights how collaborations with upstream industrial actors can complement brand-facing initiatives by enabling process standardization, quality consistency, and larger production volumes.

Taken together, these collaborations illustrate how Orange Fiber leverages brand partnerships as a strategic mechanism for market legitimation, risk-sharing, and diffusion. Rather than positioning sustainability as a purely symbolic attribute, the company embeds its material innovation within existing brand narratives and industrial infrastructures. This strategy underscores a broader implication for sustainable fiber innovation: partnerships with trusted brands are not merely marketing tools, but structural enablers that connect technological feasibility, aesthetic credibility, and consumer acceptance.

4.2 *Piñatex: From Pineapple Waste to Technical Textile Fibers*

Piñatex represents one of the most emblematic examples of plant-based leather alternatives derived from agricultural residues. Developed by the company Ananas Anam and founded by Dr. Carmen Hijosa, Piñatex is produced from pineapple leaf fibers (PALF), a by-product of pineapple cultivation that is traditionally discarded after harvesting [8, 22]. The innovation was conceived to address a dual sustainability challenge: on the one hand, reducing agricultural waste and preventing open-field burning or uncontrolled decomposition of pineapple leaves; on the other, offering an alternative to animal leather, whose environmental footprint is strongly associated with land use, water consumption, and chemically intensive tanning processes that generate toxic effluents and sludge [46].

From a circular economy perspective, Piñatex can be situated within the emerging stream of fruit-based and agro-residue textile materials, which leverage post-harvest biomass to reduce pressure on virgin resources while enabling circular value creation across the textile sector [64]. Pineapple leaves used for Piñatex require no additional land, irrigation, pesticides, or fertilizers, as they are generated within existing agricultural systems, primarily in the Philippines. According to company-reported data, each linear meter of Piñatex was estimated to avoid approximately 12 kg of CO_2 emissions, positioning the material as a climate-mitigating solution while simultaneously preventing waste accumulation at the agricultural stage [8]. This narrative contributed to Piñatex's early visibility as a flagship material within the broader discourse on sustainable and cruelty-free fashion.

Production Process and Circular Logic

The technical production process begins after pineapple fruit harvesting, when leaves—normally considered agricultural waste—are collected and bundled. Fibers are extracted through decortication, a semi-mechanized process that removes excess bio-matter while preserving the fibrous internal structure of the leaf. The resulting pineapple leaf fiber (PALF) is subsequently combined with polylactic acid (PLA), a plant-derived biopolymer obtained from renewable sources such as corn or sugar, which acts as a natural plasticizing agent to improve flexibility and mechanical performance. The Piñatex substrate is composed of approximately 80% PALF and 20% PLA [13].

The extracted fibers are transformed into a non-woven base material known as Piñafelt, which is then shipped to Europe—primarily Spain or Italy—for finishing treatments, including dyeing and coating, aimed at enhancing durability and water resistance [8]. Residual biomass generated during fiber extraction is reused as biofuel or organic fertilizer, reinforcing a closed-loop production logic [8]. Early production stages emphasize low-impact practices such as air-drying and limited water use. However, the finishing phase introduces a structural sustainability trade-off: although Piñatex is largely biodegradable (approximately 85–90%), the protective coating required to meet durability and performance standards reduces full biodegradability.

Material Performance, Design, and Applications

Piñatex is described as a versatile material compatible with conventional manufacturing processes such as cutting, sewing, and printing. It exhibits leather-like aesthetic qualities while maintaining distinct material properties, including breathability, moisture absorption, and permeability comparable to animal leather. At the same time, empirical assessments and industry feedback acknowledge lower tear resistance relative to bovine leather, which may limit adoption in product categories requiring high mechanical strength or long-term durability. Despite these constraints, Piñatex has been successfully applied in footwear, accessories, and selected apparel items, including handcrafted products positioned in premium and high-fashion segments.

Design-oriented research suggests that pineapple-based materials are particularly effective in accessories, where material storytelling and "no-waste" narratives can be translated into recognizable product forms that enhance consumer familiarity and symbolic value, even when mechanical performance is still being optimized [31]. In this respect, Piñatex illustrates how design can function as a mediating layer between material innovation and market diffusion, helping to bridge the gap between sustainability narratives and consumer-facing products.

Brand Collaborations and Market Legitimation

Brand collaborations played a crucial role in legitimizing Piñatex within the fashion system, particularly during its expansion phase. Multiple brands—including H&M, Hugo Boss, and Nike—integrated Piñatex into their collections, using brand reputation to translate material innovation into consumer trust. A particularly emblematic example is Nike's "Happy Pineapple" collection launched in 2021, which incorporated Piñatex into five iconic sneaker models (Air Force 1, Air Max 90, Air Max 95, Free Run Trail Premium, and Air Zoom Type). In this collection, Piñatex was used across multiple shoe components and visually signaled through a small pineapple logo, framing sustainability as both a material attribute and a symbolic marker of environmental commitment [38].

Beyond brand partnerships, Piñatex's credibility was further reinforced through certifications and impact reporting. Ananas Anam obtained B-Corp certification and published Impact Reports documenting environmental, social, and ethical performance, including waste reduction, resource efficiency, and job creation in rural areas. The company explicitly aligned Piñatex with the United Nations Sustainable Development Goals, contributing to eleven of the seventeen targets, and framed material innovation as a driver of both environmental and socio-economic value creation [8].

Constraints, Market Failure, and Scaling Challenges

Despite its relative technological maturity and strong symbolic positioning, Piñatex ultimately illustrates the structural fragility of many next-generation sustainable materials when confronted with real-world market conditions. In late 2025, Ananas Anam entered bankruptcy-equivalent proceedings in the United Kingdom, while its Spanish subsidiary simultaneously initiated pre-insolvency procedures. Attempts to recapitalize the company or secure acquisition failed, and declining revenues combined with rising costs undermined business continuity [49].

Industry analyses interpret Piñatex's collapse as emblematic of a broader pattern affecting next-generation materials, many of which have struggled to move beyond niche adoption toward industrial-scale commercialization. Commentators point to persistent barriers related to production costs, scalability, mechanical performance, and price competitiveness, especially in a market environment increasingly sensitive to cost–performance ratios rather than sustainability narratives alone [78]. In this framing, Piñatex's failure does not invalidate its environmental rationale but highlights a misalignment between technological promise, business model sustainability, and market readiness.

Overall, Piñatex exemplifies both the potential and the limitations of agro-residue valorization in fashion. While the material successfully demonstrated the feasibility and symbolic power of fruit-based leather alternatives, its trajectory underscores that sustainability-driven material innovation must be supported by robust industrial integration, cost-effective scaling, and durable performance to achieve long-term viability. As such, Piñatex provides a critical reference point for understanding the economic and organizational challenges that accompany the transition toward sustainable fibers and materials.

4.3 *Mylo: From Traditional Leather to an Ecological Technical Alternative*

Mylo represents one of the most advanced attempts to develop a biofabricated alternative to animal leather through biotechnology. Developed by the California-based company Bolt Threads, founded in 2009, Mylo is a mycelium-based material designed to replicate selected aesthetic and functional properties of leather while reducing environmental and ethical concerns associated with livestock farming and conventional tanning processes [16, 67]. The innovation addresses two interrelated sustainability challenges: the high environmental footprint of animal-derived leather—linked to land use, greenhouse gas emissions, and chemically intensive tanning—and the limitations of synthetic leather alternatives, which rely on fossil-based polymers and generate persistent pollution [52].

From a conceptual perspective, Mylo belongs to the broader category of next-generation bio-based materials, which seek to redesign material origins and production processes rather than improving existing fibers incrementally. Unlike waste-derived or plant-based materials, Mylo is produced through controlled biological growth, positioning biofabrication as a potentially disruptive pathway within sustainable fashion innovation [82].

Technical Production Process and Circular Logic

Mylo is produced using mycelium, the root-like network of fungi, which is cultivated under controlled indoor conditions using organic feedstocks. By regulating temperature, humidity, and growth parameters, mycelium forms dense three-dimensional structures that can be shaped into sheets resembling leather. Once the desired structure is achieved, the material undergoes dehydration and finishing processes to stabilize its form and enhance surface consistency, strength, and flexibility [16].

A key feature of Mylo's production model is the absence of chromium-based tanning and other highly hazardous chemicals commonly used in conventional leather processing. This positions Mylo within a green chemistry framework aimed at reducing toxic inputs and improving occupational and environmental safety [16]. Residual biomass generated during production is compostable, reinforcing a circular resource logic at the material level.

Unlike agro-residue–based materials, Mylo's production is not dependent on agricultural cycles or land availability. The biological growth process allows for relatively rapid material generation—on the order of weeks rather than years—suggesting potential advantages in terms of resource efficiency and production flexibility [67]. However, the process remains energy- and capital-intensive, particularly at pilot and pre-industrial scales.

Material Performance and Applications.

At the material level, Mylo is engineered to exhibit softness, elasticity, and visual uniformity comparable to premium leather. Its microstructure can be tuned to achieve different thicknesses and surface finishes, enabling applications in luxury accessories, footwear components, and experimental apparel items [16]. Early applications emphasized sensorial qualities and design experimentation rather than full functional equivalence with traditional leather across all use cases.

Industry assessments suggest that while Mylo performs well in terms of aesthetics and tactility, questions remain regarding long-term durability, abrasion resistance, and performance under intensive or prolonged use—dimensions that are particularly critical in footwear and luxury leather goods [52]. As with other next-generation materials, performance optimization remains an ongoing challenge rather than a fully resolved outcome.

Brand Collaborations and Market Positioning

Brand collaborations played a central role in legitimizing Mylo within the fashion system. Bolt Threads adopted a consortium-based diffusion model, granting selected partners—including Stella McCartney, Adidas, Kering, and Lululemon—exclusive access to the material during its early commercialization phase. High-profile product launches, such as the Frayme Mylo bag by Stella McCartney and the Stan Smith Mylo sneakers by Adidas in 2021, positioned Mylo within luxury and performance-oriented segments, framing biofabrication as compatible with premium design and brand values [67].

These collaborations functioned as credibility signals, leveraging the reputational capital of established brands to translate technological novelty into symbolic and market legitimacy. At the same time, the exclusivity of the consortium model limited diffusion beyond a small set of experimental collections, constraining broader market exposure.

Constraints and Scalability Challenges

Despite strong narrative visibility and significant financial investment, Mylo ultimately illustrates the structural challenges faced by biofabricated materials in scaling beyond pilot and showcase applications. In 2023, Bolt Threads announced the suspension of Mylo production, citing insufficient funding to support industrial-scale manufacturing under deteriorating macroeconomic conditions and shifting investor priorities [15, 50, 61].

Industry analyses highlight several interrelated barriers underlying this outcome: high production costs, limited manufacturing capacity, and difficulties in achieving cost parity and consistent quality at scale relative to conventional leather [52]. Moreover, while bio-based, Mylo is not yet fully biodegradable, prompting the company to prioritize durability and longevity over rapid degradation—an approach aligned with sustainability principles in luxury contexts but one that complicates end-of-life narratives [67].

Overall, Mylo exemplifies both the transformative potential and the fragility of biofabrication as a sustainability pathway in fashion. While biotechnology enables radical rethinking of material origins and production timelines, its long-term viability depends on overcoming substantial economic, industrial, and organizational barriers. In this sense, Mylo complements the other two cases by illustrating the outer frontier of sustainable material innovation—where technological ambition must ultimately converge with scalable business models and industrial integration to generate lasting impact.

4.4 Cross-Case Comparison

The three case studies analyzed in the previous sections illustrate distinct pathways of technical sustainable fiber and material innovation in the fashion industry, differing in technological maturity, degree of industrial integration, and proximity to established production and market practices. Orange Fiber represents a waste-to-cellulose solution that remains closely aligned with conventional textile processes; Piñatex exemplifies the valorization of agricultural residues into plant-based leather-like materials with applications in accessories and footwear; Mylo reflects the most experimental trajectory, grounded in biotechnology and biofabrication, with diffusion largely confined to selected premium and luxury partners.

Despite a shared sustainability rationale, the three cases diverge substantially in their innovation logic and implications for diffusion. Table 1 summarizes the main dimensions emerging from the cross-case comparison, highlighting differences in feedstock origin, circular principles, industrial diffusion models, and scalability challenges. A first relevant distinction concerns the degree of technological discontinuity introduced by each innovation. Orange Fiber operates relatively close to existing regenerated cellulose fiber technologies, facilitating compatibility with established textile infrastructures and reducing integration barriers. Piñatex introduces a more visible material shift by reframing agricultural residues as leather alternatives, requiring partial adaptation of production and finishing processes. Mylo represents the most radical departure, relying on biological growth and biofabrication rather than traditional fiber processing, thereby challenging established industrial paradigms [54, 82].

These differences are mirrored in the industrial diffusion strategies adopted across cases. Orange Fiber follows a partnership-based scaling model, embedding its innovation within existing textile value chains through collaborations with fiber producers

Table 1 Comparison between the three cases

Dimension	Orange fiber	Piñatex	Mylo
Technical innovation logic	Waste-to-cellulose textile fiber	Agro-residue–based leather alternative	Biofabricated leather alternative
Feedstock	Citrus processing waste	Pineapple leaves (agricultural residue)	Mycelium (cultivated biomass)
Circular principle	Waste valorization within existing textile chains	Use of agricultural residues without additional land use	Substitution of animal leather via biological growth
Main applications	Apparel textiles (premium positioning)	Footwear and accessories (leather-like)	Premium and luxury experimental products
Industrial diffusion model	Partnerships with established textile and fashion players	Brand collaborations and certified impact-oriented scaling	Selective partnerships with major brands
Main scalability challenges	Process optimization, cost, and quality consistency	Material resistance and finishing trade-offs	High costs, limited scale, technological uncertainty

Source Authors' own elaboration

and fashion brands. Piñatex relies on a combination of brand collaborations, certifications, and impact reporting to build credibility and market visibility, consistent with diffusion patterns observed in sustainability-oriented material innovations [17]. By contrast, Mylo adopts a highly selective partnership approach typical of early-stage biomaterials, where adoption is concentrated among large brands capable of absorbing higher costs, technological uncertainty, and reputational risk [23, 73].

Scalability challenges also diverge significantly across the three cases. For Orange Fiber, the main constraints concern process optimization, cost efficiency, and the ability to ensure consistent quality at scale within waste-based supply chains. Piñatex faces trade-offs between sustainability objectives and material performance, particularly in applications requiring high mechanical resistance and durability. Mylo encounters the most structural barriers, including high production costs, limited manufacturing scale, and unresolved questions related to long-term durability and industrial viability [7, 82]. Taken together, these patterns indicate that sustainable materials follow uneven diffusion trajectories shaped less by sustainability intent per se than by technological maturity, supply-chain readiness, and the capacity to integrate innovation within existing industrial ecosystems.

From a market and adoption perspective, the cases also suggest that proximity to established material categories plays a critical role in shaping diffusion potential. Innovations that remain closer to conventional textile standards—such as regenerated cellulose fibers—are more readily integrated into apparel production, while materials that challenge entrenched categories, such as leather alternatives or biofabricated materials, face higher barriers related to performance expectations, cost structures, and industrial compatibility. Prior research indicates that such technological distance can influence acceptance dynamics in fashion, particularly in contexts where durability, tactility, and embodied experience are central to value perception [1, 4].

Overall, the cross-case comparison reinforces a core implication of this chapter: material innovation alone is insufficient to ensure diffusion and impact (Table 1). The transition toward sustainable fibers and materials depends on the alignment of technological feasibility, industrial integration, and credible value propositions capable of translating sustainability attributes into functional, aesthetic, and economic value. This insight is particularly salient in luxury and premium fashion contexts, where sustainability must coexist with high expectations regarding quality, durability, and sensory legitimacy [25, 40, 53].

5 Discussion

This chapter conceptualizes eco-innovation in technical innovation textiles as a structural response to the environmental crisis of the fashion system, rather than as a peripheral or symbolic adjustment. One of its central contributions lies in positioning fibers as a critical lever in the transition toward circular models, coherently with the circular economy perspective that replaces end-of-life logic with restorative processes, renewable energy, and the elimination of toxic chemistry [27]. Within this framework, recycling—and particularly upcycling—emerge as key mechanisms for reducing carbon footprints, water consumption, and waste generation. Across the chapter and the three case studies, it becomes evident that fiber innovation is not an “add-on” to sustainability; rather, it is a domain in which a substantial share of fashion’s environmental impact is concentrated, given that material production and processing account for a large portion of the industry’s footprint [7].

Furthermore, a distinctive element of this chapter is the integration of sustainable luxury into the discussion of materials, highlighting a fundamental tension. While prior research warns that sustainability cues may undermine perceived luxury quality and reduce purchase intentions [1], other studies suggest that consumers—especially younger cohorts—expect luxury brands to uphold higher ethical and environmental standards, and that sustainability-oriented communication can reposition luxury as simultaneously “gold and green” [9]. Moreover, the chapter argues that quality and durability, which are intrinsic to luxury, may render certain luxury goods inherently more sustainable than mass-market products, due to longer lifecycles and more responsible disposal practices [4, 24, 60]. Navigating this tension requires avoiding two symmetrical pitfalls: reducing sustainability to a narrative that weakens luxury codes, or reducing luxury to declared sustainability without material outcomes—an issue consistent with evidence that increased reporting does not necessarily correspond to substantive progress [74]. The three case studies operationalize these arguments by illustrating complementary innovation trajectories. Orange Fiber exemplifies a waste-to-resource model based on the valorization of citrus by-products and strong industrial partnerships compatible with luxury positioning; Piñatex demonstrates how agricultural residues, such as pineapple leaves, can be transformed into certified leather alternatives for footwear and accessories; and Mylo represents the biotech frontier, where mycelium-derived materials allow control over material properties but face typical challenges related to scaling and cost. Taken together, these cases indicate that the diffusion of innovative fibers depends on supply-chain collaboration, the ability to manage uncertainty, and communication strategies that reduce perceived risk while emphasizing tangible consumer benefits and innovativeness [23, 40, 73]. From a managerial perspective, the chapter confirms that green strategies can generate environmental and economic advantages through waste reduction, minimization of hazardous substances, and reverse logistics, thereby enhancing performance and reputation [51, 57], while also acknowledging that such benefits are not automatic and depend on coordination, timing, quality, and operational capabilities within complex supply chains. In this context, product and process innovation in fibers emerges as a high-leverage form of green innovation, as it intervenes directly in the material foundation of fashion products rather than relying solely on communication or downstream practices. The growing policy pressure and acceleration of sustainability targets [30, 35, 37], combined with projected growth in textile volumes [30], further reinforce the strategic relevance of this pathway. Nevertheless, the chapter also identifies persistent barriers to scalability, including technological complexity and output quality constraints—particularly salient in luxury contexts—high cost structures, limited supply-chain integration for collection and preprocessing, and challenges related to consumer adoption, as novelty can both enhance evaluations and generate scepticism when innovations are perceived as uncertain or inappropriate for close-to-skin products [7, 20, 58]. Overall, innovative fibers emerge as a promising response to fashion's sustainability challenges, but their widespread adoption depends on enabling conditions such as process innovation, coordinated supply chains, credible certifications and labeling systems, and communication capable of translating material innovation into perceived value.

6 Conclusion

This chapter examined eco-innovation in technical textiles by positioning **sustainable fibers** as a key lever to transform the fashion system toward more circular and responsible models. Consistent with the abstract submitted to the editor, the central argument is that reducing fashion's environmental impact—strongly linked to water use, production chemistry, textile waste, and emissions—requires changes in both industrial models (moving beyond linear logic) and material choices [21, 27, 28, 59].

The chapter also emphasized how fast fashion has intensified environmental and social pressures, strengthening the urgency of strategies such as reducing overproduction, extending product lifespans, improving circularity, and developing material alternatives [33, 35]. In this scenario, fiber innovation—through recycling, upcycling, and biomaterials—emerges as a frontier that connects innovation, circularity, and consumer culture.

The case studies (Orange Fiber, Piñatex, Mylo) demonstrate that eco-innovation can take different forms (waste-to-fiber, leather alternatives, biotech), but converge on one point: sustainability depends on the ability to transform waste and renewable resources into high-performing materials that can be integrated into supply chains and align with consumer expectations, including within the luxury segment. Luxury brands may play a decisive role by accelerating adoption and legitimization of innovative fibers, yet they must manage the tension between luxury codes and sustainability signals, avoiding negative effects on perceived quality [1] while leveraging "gold and green" narratives [9].

In conclusion, sustainable fibers represent a promising frontier for the future of fashion, but their scalability depends on overcoming technological, economic, and organizational barriers, and on building a collaborative ecosystem among firms, suppliers, policymakers, and consumers. By placing fibers at the intersection of innovation, circularity, and consumer culture, this chapter argues that eco-innovation in textiles is not merely a trend, but a strategic infrastructure for reimagining the future of the fashion system.

References

1. Achabou MA, Dekhili S (2013) Luxury and sustainable development: is there a match? J Bus Res 66(10):1896–1903
2. Adıgüzel F, Donato C (2021) Proud to be sustainable: upcycled versus recycled luxury products. J Bus Res 130:137–146
3. Aishwariya S (2020) Textiles from orange peel waste. Sci Technol Dev J 23(2)
4. Amatulli C, De Angelis M, Donato C (2021) The atypicality of sustainable luxury products. Psychol Mark 38(11):1990–2005
5. Amatulli C, De Angelis M, Pino G, Guido G (2020) An investigation of unsustainable luxury: how guilt drives negative word-of-mouth. Int J Res Mark 37:821–836
6. Ambec S, Lanoie P (2008) Does it pay to be green? A systematic overview. Acad Manag Perspect 22(4):45–62
7. Amed, I., André, S., Balchandani, A., Berg, A., & Rölkens, F. (2022). The state of fashion 2023: holding onto growth as global clouds gather. McKinsey & Company. Retrieved from https://www.mckinsey.com/industries/retail/our-insights/state-of-fashion. Accessed on 3 Nov 2025
8. Ananas Anam (2023) Piñatex® description. Retrieved from: https://www.ananas-anam.com/
9. De Angelis M, Adıgüzel F, Amatulli C (2017) The role of design similarity in consumers' evaluation of new green products: an investigation of luxury fashion brands. J Clean Prod 141:1515–1527
10. Arena E, Valenti G (2022) Orange fiber: Dal Laboratorio Alla Moda. Chimica & Tessile Conciario la Chimica e L'industria Online Anno VI 4. http://dx.medra.org/10.17374/CI.2022.104.4.26
11. Aus R, Moora H, Vihma M, Unt R, Kiisa M, Kapur S (2021) Designing for circular fashion: integrating upcycling into conventional garment manufacturing processes. Fash Text 8:1–18
12. Baier D, Rausch TM, Wagner TF (2020) The drivers of sustainable apparel and sportswear consumption: a segmented Kano perspective. Sustainability 12(7):2788
13. Banik S, Nag D, Debnath S (2011) Utilization of pineapple leaf agro-waste for extraction of fibre and the residual biomass for vermicomposting. Indian J Fibre Text Res 36(2):172
14. Benigni A (2025) Orange Fiber: il tessile sostenibile e circolare delle arance siciliane diventa moda internazionale. Retrieved from https://circulareconomyforfood.eu/orange-fiber/. Accessed on 2 Jan 2026
15. Bittau L (2023) Frenano ancora i materiali next-gen. Flop per la pelle vegetale Mylo. Retrieved from https://www.pambianconews.com/2023/07/04/frenano-ancora-i-materiali-next-gen-in-stand-by-la-pelle-vegetale-mylo-379388/
16. Bolt Threads (2020) Mylo™ mycelium leather: technical overview and material information. Bolt Threads. https://mylo-unleather.com/
17. Braungart M, McDonough W, Bollinger A (2007) Cradle-to-cradle design: creating healthy emissions–a strategy for eco-effective product and system design. J Clean Prod 15(13–14):1337–1348
18. Bridgens B, Powell M, Farmer G, Walsh C, Reed E, Royapoor M et al (2018) Creative upcycling: reconnecting people, materials and place through making. J Clean Prod 189:145–154
19. Caniato F, Caridi M, Crippa L, Moretto A (2012) Environmental sustainability in fashion supply chains: an exploratory case based research. Int J Prod Econ 135(2):659–670
20. Carpenter GS, Glazer R, Nakamoto K (1994) Meaningful brands from meaningless differentiation: the dependence on irrelevant attributes. J Mark Res 31(3):339–350
21. Cho (2021) Why Fashion Needs to Be More Sustainable. News from the Columbia Climate School. Available at https://news.climate.columbia.edu/2021/06/10/why-fashion-needs-to-be-more-sustainable/
22. Debnath S (2016) Natural fibres for sustainable development in fashion industry. In: Sustainable fibres for fashion industry: volume 1. Springer Singapore, Singapore, pp 89–108

23. D'Adamo I, Lupi G, Morone P, Settembre-Blundo D (2022) Towards the circular economy in the fashion industry: the second-hand market as a best practice of sustainable responsibility for businesses and consumers. Environ Sci Pollut Res 29(31):46620–46633
24. D'Aniello A, Amatulli C, De Angelis M, Pozharliev R (2025). Principles, cases, and key challenges. In: The Routledge companion to marketing and sustainability
25. D'Aniello A, Levy ACM, Amatulli C, De Angelis M, Pozharliev R (2025) Aesthetics and sustainable luxury consumption: evidence from scientific literature. In: Case studies and consumer neuroscience. Springer Nature
26. Eisenhardt KM, Graebner ME (2007) Theory building from cases: opportunities and challenges. Acad Manag J 50(1):25–32
27. Ellen MacArthur Foundation (2013) Towards the circular economy Vol. 1: an economic and business rationale for an accelerated transition. Retrieved from https://ellenmacarthurfoundation.org/towards-the-circular-economy-vol-1-an-economic-and-business-rationale-for-an. Accessed on 1 Dec 2025
28. Ellen MacArthur Foundation (2017) A New Textiles Economy: redesigning fashion's future. Available at https://www.ellenmacarthurfoundation.org/a-new-textiles-economy
29. Eppinger E (2022) Recycling technologies for enabling sustainability transitions of the fashion industry: status quo and avenues for increasing post-consumer waste recycling. Sustain Sci Pract Policy 18(1):114–128
30. European Commission (2022) EU strategy for sustainable and circular textiles. Communication from the Commission to the European Parliament, the Council, the European Economic and Social Committee and the Committee of the Regions. Retrieved from https://eur-lex.europa.eu/legal-content/EN/TXT/HTML/?uri=CELEX:52022DC0141. Accessed on 1 Dec 2025
31. Falsafi Z (2024) PinaClutch-no waste! In: International textile and apparel association annual conference proceedings, vol 80, No. 1. Iowa State University Digital Press
32. Fletcher K (2008) Sustainable fashion and clothing. Design Journeys Earthscan Malta 10:9781315857930
33. Garcia-Ortega B, Galan-Cubillo J, Llorens-Montes FJ, de-Miguel-Molina B (2023) Sufficient consumption as a missing link toward sustainability: the case of fast fashion. J Clean Prod 399:136678
34. Gupta V, Arora M, Minhas J (2020) Innovating opportunities for fashion brands by using textile waste for better fashion. In: Recycling from waste in fashion and textiles: a sustainable and circular economic approach, pp 101–121
35. Hedrich S, Janmark J, Langguth N, Magnus K, Strand M (2022) Scaling textile recycling in Europe-turning waste into value. McKinsey & Company. Retrieved from https://www.mckinsey.com/industries/retail/our-insights/scaling-textile-recycling-in-europe-turning-waste-into-value. Accessed on 30 Nov 2025
36. Henninger CE, Alevizou PJ, Oates CJ (2016) What is sustainable fashion? J Fashion Mark Manag Int J 20(4):400–416
37. Hieronimus S, Leke A, Ngai J, Rodeia P, Hilton Segel L, Smit S (2023) McKinsey and the world economic forum 2023. McKinsey & Company. Retrieved from https://www.mckinsey.com/featured-insights/world-economic-forum/overview. Accessed on 30 Nov 2025
38. Ho S (2021) Nike partners with Pinatex for 'Happy Pineapple' vegan leather sneaker collection. Retrieved from https://www.greenqueen.com.hk/nike-vegan-pineapple-leather-sneakers/. Accessed on 20 Dec 2025
39. Hofstetter JS, De Marchi V, Sarkis J, Govindan K, Klassen R, Ometto AR et al (2021) From sustainable global value chains to circular economy—different silos, different perspectives, but many opportunities to build bridges. Circ Econ Sustain 1(1):21–47
40. Hubert M, Florack A, Gattringer R, Eberhardt T, Enkel E, Kenning P (2017) Flag up!-flagship products as important drivers of perceived brand innovativeness. J Bus Res 71:154–163
41. Hwang CG, Lee YA, Diddi S (2015) Generation Y's moral obligation and purchase intentions for organic, fair-trade, and recycled apparel products. Int J Fash Des Technol Educ 8(2):97–107
42. Ivanov D, Dolgui A (2020) Viability of intertwined supply networks: extending the supply chain resilience angles towards survivability. A position paper motivated by COVID-19 outbreak. Int J Prod Res 58(10):2904–2915

43. Jain S (2019) Factors affecting sustainable luxury purchase behavior: a conceptual framework. J Int Consum Mark 31(2):130–146
44. Jalil MA, Repon MR, Jurkonienė S, Haji A, Hussain SZ, Shukhratov S (2024) Valorization of pineapple leaves: effective conversion of agro waste to textile materials. Energy Sci Eng 12(6):2426–2434
45. Kamleitner B, Thürridl C, Martin BA (2019) A Cinderella story: how past identity salience boosts demand for repurposed products. J Mark 83(6):76–92
46. Kefale GY, Kebede ZT, Birlie AA (2023) A systematic review on potential bio leather substitute for natural leather. J Eng 2023(1):1629174
47. Khan A, Awais M, Mohsin M, Khan A, Cheema K (2024) Sustainable yarns and fabrics from tri-blends of banana, cotton and tencel fibres for textile applications. J Clean Prod 436:140545
48. Kozlowski A, Searcy C, Bardecki M (2019) The reDesign canvas: fashion design as a tool for sustainability. J Clean Prod 234:1214–1228
49. LaConceria (2025) Fallisce Piñatex, la pelle che non era pelle e nemmeno business. LaConceria—Il portale dell'area pelle. https://www.laconceria.it/sostenibilita/fallisce-pinatex-la-pelle-che-non-era-pelle-e-nemmeno-business/
50. LaConceria (2023) Mylo si schianta: senza soldi chiude l'alternativa veg alla pelle. Retrieved from https://www.laconceria.it/finanza/mylo-si-schianta-senza-soldi-chiude-lalternativa-veg-alla-pelle/
51. Lee KH, Kim JW (2011) Integrating suppliers into green product innovation development: an empirical case study in the semiconductor industry. Bus strategy Environ 20(8):527–538
52. Lineapelle Mag (2023) La favoletta senza lieto fine di Mylo l'Alternativo. Retrieved from https://magazine.lineapelle-fair.it/green-corner/la-favoletta-senza-lieto-fine-di-mylo-lalternativo/
53. Luchs MG, Naylor RW, Irwin JR, Raghunathan R (2010) The sustainability liability: potential negative effects of ethicality on product preference. J Mark 74(5):18–31
54. McKinsey (2020) Fashion on climate. How the fashion industry can reduce its carbon footprint. Available at https://www.mckinsey.com/industries/retail/our-insights/fashion-on-climate. Accessed on 23 Dec 2025
55. McKinsey (2025) The state of fashion 2026: when the rules change. Available at https://www.mckinsey.com/industries/retail/our-insights/state-of-fashion. Accessed on 23 Dec 2025
56. McNeill L, Moore R (2015) Sustainable fashion consumption and the fast fashion conundrum: fashionable consumers and attitudes to sustainability in clothing choice. Int J Consum Stud 39(3):212–222
57. Miroshnychenko I, Barontini R, Testa F (2017) Green practices and financial performance: a global outlook. J Cleaner Prod 147:340–351
58. Mukherjee A, Hoyer WD (2001) The effect of novel attributes on product evaluation. J Consum Res 28(3):462–472
59. Niinimäki K, Peters G, Dahlbo H, Perry P, Rissanen T, Gwilt A (2020) The environmental price of fast fashion. Nat Rev Earth Environ 1(4):189–200
60. Notaro S, Paletto A (2021) Consumers' preferences, attitudes and willingness to pay for bio-textile in wood fibers. J Retail Consum Serv 58:102304
61. Nssmag.com (2023) L'azienda che produce di pelle vegana per Stella McCartney e adidas sta fallendo. Retrieved from: https://www.nssmag.com/it/fashion/33696/mylo-pelle-vegana-stella-mccartney
62. Orange Fiber Website (n.d..): https://orangefiber.it/it/
63. Park HJ, Lin LM (2018) Exploring attitude–behavior gap in sustainable consumption: comparison of recycled and upcycled fashion products. J Bus Res 117:623–628
64. Pešić M, Nemeša I, Buhonka N, Bozokifryt V (2023) Fruit based sustainable textile materials. In: 9th International joint conference on environmental and light industry technologies, pp 91–99
65. Plakantonaki S, Kiskira K, Zacharopoulos N, Belessi V, Sfyroera E, Priniotakis G, Athanasekou C (2024) Investigating the routes to produce cellulose fibers from agro-waste: an upcycling process. ChemEngineering 8(6):112

66. Rana P, Sethi S (2025) Utilising agro-waste in textile industry: advancing sustainability with a year-round waste cycle. Mater Circ Econ 7(1):30
67. van Rooyen A (2024) Sustainable fashion tech innovation: a case study of bolt threads' redefinition of fashion systems, design activism, and social entrepreneurship. DISCERN: Int J Des Soc Change Sustain Innov Entrepreneur 5(2):1–13
68. Savio L, Pennacchio R, Patrucco A, Manni V, Bosia D (2022) Natural fibre insulation materials: use of textile and agri-food waste in a circular economy perspective. Mater Circ Econ 4(1):6
69. Sigaard AS, Laitala K (2023) Natural and sustainable? Consumers' Textile Fiber Preferences Fibers 11(2):12
70. Siqueira MU, Contin B, Fernandes PRB, Ruschel-Soares R, Siqueira PU, Baruque-Ramos J (2022) Brazilian agro-industrial wastes as potential textile and other raw materials: a sustainable approach. Mater Circ Econ 4(1):9
71. Sproles GB, Kendall EL (1986) A methodology for profiling consumers' decision-making styles. J Consum Aff 20(2):267–279
72. Tessuti di Sondrio (n.d.). Tessuti in Orange Fiber®: Innovazione e manifattura tessile italiana. Retrieved from: https://www.tessutidisondrio.it/i-tessuti/le-collezioni/heart-for-earth/orange-fiber/
73. Thanasopon B, Papadopoulos T, Vidgen R (2016) The role of openness in the fuzzy front-end of service innovation. Technovation 47:32–46
74. The Business of Fashion (2022) The BoF Sustainability Index 2022. Retrieved from https://insights.businessoffashion.com/products/the-bof-sustainability-index-2022. Accessed on 18 Dec 2025
75. The Business of Fashion (2025). The Best of BoF 2025: luxury scandals and the retreat of "Sustainable Fashion". The Business of Fashion. Retrieved from https://www.businessoffashion.com/articles/sustainability/the-best-of-bof-2025-luxury-scandals-and-the-retreat-of-sustainable-fashion/. Accessed on 19 Dec 2025
76. Todeschini BV, Cortimiglia MN, Callegaro-de-Menezes D, Ghezzi A (2017) Innovative and sustainable business models in the fashion industry: entrepreneurial drivers, opportunities, and challenges. Bus Horiz 60(6):759–770
77. Torrisi C, Carrabino R, Santonocito A, Matarazzo A, Mertoli F (2022) Industrial symbiosis in the production of sustainable textiles: orange fiber SRL. P-ESEM, 995
78. Viti M (2025) Un ko dopo l'altro per i materiali Next Gen: l'ultimo è Piñatex. Lineapelle Magazine—Green Corner. https://magazine.lineapelle-fair.it/green-corner/un-ko-dopo-laltro-per-i-materiali-next-gen-lultimo-e-pinatex
79. Walsh PR, Dodds R (2017) Measuring the choice of environmental sustainability strategies in creating a competitive advantage. Bus Strateg Environ 26(5):672–687
80. Wilson M (2016) When creative consumers go green: understanding consumer upcycling. J Prod Brand Manag 25(4):394–399
81. Winterich KP, Nenkov GY, Gonzales GE (2019) Knowing what it makes: how product transformation salience increases recycling. J Mark 83(4):21–37
82. Wood J (2019) Bioinspiration in fashion—a review. Biomimetics 4(1):16
83. Yin RK (2014) Case study research: design and methods (applied social research methods)
84. Zhang Z, Huang J, Yao Y, Peters G, Macdonald B, La Rosa AD, Scherer L (2023) Environmental impacts of cotton and opportunities for improvement. Nat Rev Earth Environm 4(10):703–715

Correction to: Ultraviolet Protective Textiles: Mechanisms, Material Parameters and Sustainable Approaches

Sanjida Sultana, Tanzima Zoha Chowdhury, Sushama Saha Swati, Zinia Yasmin, and Mahfuza Pervin

Correction to:
Chapter 5 in: S. S. Muthu (ed.), *Technical Textiles for Sustainability*, Sustainable Textiles: Production, Processing, Manufacturing & Chemistry, https://doi.org/10.1007/978-3-032-29974-1_5

In the original version of the book, the misspelt co-author name "Sushoma Shaha Swati" has been changed to read as "Sushama Saha Swati" in Chapter 5. The erratum chapter and the book have been updated with the change.

The updated version of this chapter can be found at
https://doi.org/10.1007/978-3-032-29974-1_5

S. S. Muthu (ed.), *Technical Textiles for Sustainability*, Sustainable Textiles: Production, Processing, Manufacturing & Chemistry, https://doi.org/10.1007/978-3-032-29974-1_7

MIX
Papier aus verantwortungsvollen Quellen
Paper from responsible sources
FSC® C105338

If you have any concerns about our products,
you can contact us on
ProductSafety@springernature.com

In case Publisher is established outside the EU,
the EU authorized representative is:
Springer Nature Customer Service Center GmbH
Europaplatz 3, 69115 Heidelberg, Germany

Printed by Libri Plureos GmbH
in Hamburg, Germany